Fundamentals of Analysis in Physics

Authored by

Masatoshi Kajita

National Institute of Information and Communications Technology,
Koganei, Tokyo, Japan

Fundamentals of Analysis in Physics

Author: Masatoshi Kajita

ISBN (Online): 978-981-5049-10-7

ISBN (Print): 978-981-5049-11-4

ISBN (Paperback): 978-981-5049-12-1

©2022, Bentham Books imprint.

Published by Bentham Science Publishers Pte. Ltd. Singapore. All Rights Reserved.

need for a court order if at any point you breach any terms of this License Agreement. In no event will any delay or failure by Bentham Science Publishers in enforcing your compliance with this License Agreement constitute a waiver of any of its rights.

3. You acknowledge that you have read this License Agreement, and agree to be bound by its terms and conditions. To the extent that any other terms and conditions presented on any website of Bentham Science Publishers conflict with, or are inconsistent with, the terms and conditions set out in this License Agreement, you acknowledge that the terms and conditions set out in this License Agreement shall prevail.

Bentham Science Publishers Pte. Ltd.
80 Robinson Road #02-00
Singapore 068898
Singapore
Email: subscriptions@benthamscience.net

CONTENTS

PREFACE

This book summarizes the analysis in whole fields of physics without using the special functions, targeting college/bachelor students. Many beginners feel that it is difficult to learn each field of physics (classical mechanics, electromagnetism, quantum mechanics, relativistic quantum mechanics, statistic mechanics) in detail separately. It would be preferable to learn the whole fields as quick as possible and have a simple imagination about the relation between different fields. After learning the position of each field in the physics, it becomes easier to learn detailed parts of each field. In this book, the fundamental of mathematical treatments are introduced, which are important for the analysis in physics but not familiar to all the readers. The fundamental of the analysis in each field of physics are summarized afterwards. The estimation of measurement uncertainty is also introduced. The important points of whole physics are summarized within 150 pages.

It would be my great pleasure, if this book can help college students to understand the fundamental of physics. And I believe, it is also useful for researchers to develop new research fields.

CONSENT FOR PUBLICATION

Not applicable.

CONFLICT OF INTEREST

The author confirms that there is no conflict of interest.

ACKNOWLEDGEMENTS

The research activity of the author is supported by a Grant-in-Aid for Scientific Research (B) (Grant No. JP 17H02881 and JP20H01920) and a Grant-in-Aid for Scientific Research (C) (Grant Nos. JP 17K06483 and 16K05500) from the Japan Society for the Promotion of Science (JSPS). I want to thank Editage (www.editage.com) for English language editing.

Masatoshi Kajita
National Institute of Information and Communications
Technology, Koganei,
Tokyo, Japan

CHAPTER 1

Fundamentals of Mathematical Treatments

Abstract: Physical phenomena can be understood by solving equations that lead to physical laws. The first objective of this chapter is to solve certain equations that are required for physical analysis. First, an iterative solution of the equation is introduced. Using this approach, the numerical solution of an equation $f(x) = 0$ can be obtained also when the function $f(x)$ is too complicated for the solution to be obtained as an explicit formula.

Many physical equations can be expressed using differential and integral mathematical representations, which might not be familiar to all college students. The fundamental concepts of the differential and integral were introduced. Several fundamental mathematical formulae are reviewed.

The second objective is to solve the differential equations that are required for the physical analysis. First, some solutions of simple differential equations given by explicit formulas are introduced, which are important for their physical interpretation. However, the equations for technical use are generally too complicated. Several methods for obtaining numerical solutions are introduced, which are useful for analyzing motion orbits. There is also a phenomenon that cannot be predicted by solving equations, which is called "chaos."

Finally, the fundamentals of the eigenvalues of matrices are introduced, which are important for understanding quantum mechanics. This chapter was prepared for undergraduate students who are not familiar with differential and integral calculus and matrices.

Keywords: Iterative solution, Differential, Partial derivative, Integral, Taylor expansion, Euler method, Middle point method, Runge-Kutta method, Chaos, Lyapunov exponent, matrix, Determinant, Eigen value, Eigen vector.

1.1. INTRODUCTION

Many physical laws have been established based on equations, and physical phenomena are predicted by solving them. It is not always a simple task to solve these equations. For example, the Newtonian law of universal gravitation is given by the simple formula. However, it is not easy to analyze the motion of astronomical bodies when considering the interaction between more than three bodies.

The technical development of any analysis plays an important role in physics. For some cases, the simplification of equations makes it possible to obtain a solution that is expressed as simple formulas, which can provide new physical insights. However, numerical calculations make it possible to obtain the solutions of equations which cannot be expressed as a rigorous formula. The technical development of numerical calculations is important to minimize the error of solutions and the calculation time.

Physicists and engineers have many techniques for obtaining reliable solutions to equations. Different kinds of approximations have been developed in the field of theoretical physics, which are not acceptable for mathematicians. This chapter summarizes the fundamentals of solving the equations analytically and numerically.

Reference [1] seems to be useful to learn the fundamental of mathematics more in detailed for students, who are interested with physical analysis. Reference [2] seems to be readable also for high school students [2].

1.2. ITERATIVE SOLUTION OF EQUATIONS

There are many cases for which the solution x_0 that satisfies the following equation must be obtained:

$$f(x_0) = 0 \qquad (1.2.1)$$

Considering the following simple equations, the solutions are given by:

$$f(x) = ax + b \rightarrow x_0 = -\frac{b}{a} \qquad (1.2.2)$$

$$f(x) = ax^2 + bx + c \rightarrow x_0 = \frac{-b \pm \sqrt{b^2 - 4ac}}{2a} \qquad (1.2.3)$$

In many other cases, the solutions cannot be described using simple formulas. Therefore, solutions are often obtained using an iterative method. When $f(x_a) > 0$ $f(x_b) < 0$ with $x_a < x_b$, the solution x_0 is $x_a < x_0 < x_b$. Then we calculate $f(x_c)$ with $x_a < x_c < x_b$. If $f(x_c) > 0$, $x_c < x_0 < x_b$ is obtained. By repeating this calculation, the possible region of x_0 becomes narrower. It seems useful to choose x_c as follows:

$$x_c = \frac{|f(x_b)|x_a + |f(x_a)|x_b}{|f(x_b)| + |f(x_a)|} \tag{1.2.4}$$

because x_c is expected to be close to x_0, assuming that $f(x)$ is approximately linear in a limited region of x. For example, we obtain the solution of $f(x) = \cos(x) - x = 0$ using the following procedure.

$f(0) = 1 > 0, f(1) = -0.46 < 0,$ $0 < x_0 < 1$
$f(0.685) = 0.089 > 0, f(1) = -0.46\ < 0,$ $0.685 < x_0 < 1$
$f(0.736) = 0.005 > 0, f(0.75) = -0.018 < 0,$ $0.736 < x_0 < 0.75$
$f(0.738) = 0.00018 > 0, f(0.74) = -0.0015 < 0,$ $0738 < x_0 < 0.74$
$f(0.739) = 0.00014 > 0, f(0.7395) = -0.00069 < 0,$ $0.739 < x_0 < 0.7395$

The value of x_0 was obtained with an uncertainty below 0.03 %, and this uncertainty was further reduced by continuing the calculation with a narrower region of x. Biological   evolution is the search for a solution for survival for a change of circumstances, which is similar to the iterative method. Evolution is possible in many different directions. Species with the appropriate adaptions (advantageous traits for survival in a new circumstance) can survive, but those with disadvantageous traits will not. Therefore, biological species with desirable adaptions dominate for a long period after the change in circumstance.

The iterative method cannot be used for discontinuous functions (*e.g.*, $1/x$ at $x = 0$). Note also that the iterative method cannot be used for cases without solutions. There may also be multiple solutions, and the iterative method should be performed with a limited region of x.

1.3. DIFFERENTIAL AND INTEGRAL EQUATIONS

Before introducing differential equations, we summarize the characteristics of the differential and integral. Considering a function $y(x)$, the differential of y is defined by (see Fig. **1.1**):

$$\frac{dy(x)}{dx} = \lim_{h \to 0} \frac{y(x+h)-y(x)}{h} \tag{1.3.1}$$

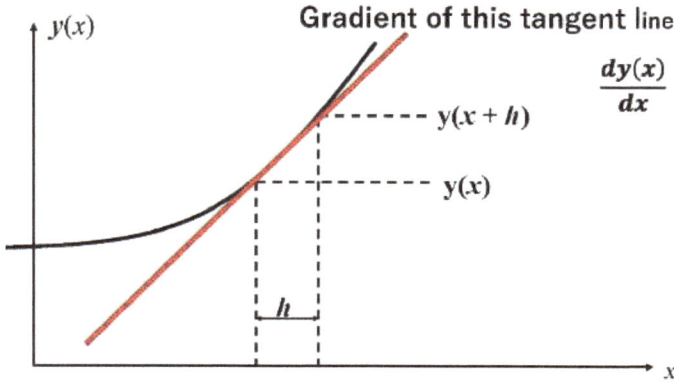

Fig. (1.1). The concept of the differential, which is the gradient of the tangent line.

The following relations are important to understand the analysis shown in other chapters.

$$\frac{d[c_y y + c_z z]}{dx} = c_y \frac{dy}{dx} + c_z \frac{dz}{dx} \tag{1.3.2}$$

$$\frac{d[yz]}{dx} = z \frac{dy}{dx} + y \frac{dz}{dx} \tag{1.3.3}$$

$$\frac{dx}{dy} = \frac{1}{\left(\frac{dy}{dx}\right)} \tag{1.3.4}$$

$$\frac{dz}{dx} = \frac{dz}{dy}\frac{dy}{dx} \tag{1.3.5}$$

$$\frac{dx^n}{dx} = nx^{n-1} \tag{1.3.6}$$

$$\frac{da^x}{dx} = \frac{de^{x\ln(a)}}{dx} = \ln(a)\,e^{x\ln(a)} = \ln(a)a^x \quad \ln(a) = \log_e(a)$$

$$e = \lim_{N \to \infty}\left(1 + \frac{1}{N}\right)^N \quad e^x = \lim_{N \to \infty}\left(1 + \frac{x}{N}\right)^N \quad \frac{de^x}{dx} = e^x$$

$$(e^x \text{ is often described by } \exp(x)) \tag{1.3.7}$$

$$\frac{d[\ln x]}{dx} = \frac{1}{x} \tag{1.3.8}$$

$$\frac{d\sin(x)}{dx} = \cos(x), \quad \frac{d\cos(x)}{dx} = -\sin(x)$$

using $e^{ix} = \cos x + i \sin x$ (Euler's theorem)

$$\frac{d(e^{ix})}{dx} = ie^{ix} = -\sin x + i \cos x \qquad (1.3.9)$$

When a function z is a function of x and y, the partial derivative $\frac{\partial z}{\partial x}$ is the differential when another parameter, y, is constant. When y is also a function of x, the total derivative is given by:

$$\frac{dz}{dx} = \frac{\partial z}{\partial x} + \frac{\partial z}{\partial y}\frac{dy}{dx} \qquad (1.3.10)$$

The primitive function $F(x)$ of a function $f(x)$ is a function given by:

$$\frac{dF(x)}{dx} = f(x) \qquad (1.3.11)$$

Then $F(x)$ is described by:

$$F(x) = \int f(x)dx \qquad (1.3.12)$$

The definite integral at $a \leq x \leq b$ is given by

$$\int_a^b f(x)\,dx = F(b) - F(a) = \lim_{N \to \infty} \sum_{n=0}^{N} f(a + n\Delta x)(\Delta x) \quad \Delta x = \frac{b-a}{N} \qquad (1.3.13)$$

which shows the area of the graph of $f(x)$ at $a \leq x \leq b$ as shown in Fig. (**1.2**).

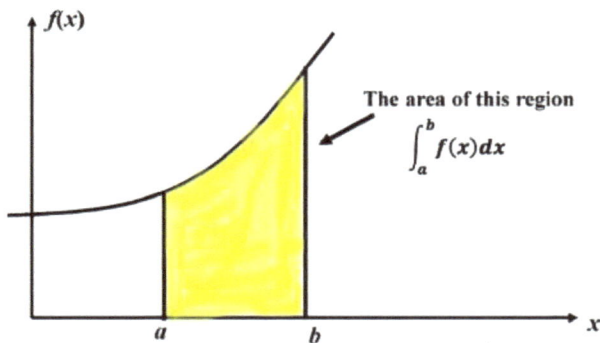

Fig. (1.2). The concept of the definite integral as the area of a region of a graph.

Using eq. (1.3.13), eq. (1.3.11) is derived from:

$$\lim_{\Delta x \to 0} \frac{F(b) - F(b - \Delta x)}{\Delta x} = f(b) \qquad (1.3.14)$$

The following relations with respect to the integral are important for understanding the analysis presented in this book.

$$\int [c_1 f_1(x) + c_2 f_2(x)] dx = c_1 \int f_1(x) dx + c_2 \int f_2(x) dx \qquad (1.3.15)$$

$$\int f(y) dx = \int f(y) \frac{dx}{dy} dy \qquad (1.3.16)$$

$$\int f(x) g(x) dx = f(x) G(x) - \int \left(\frac{df(x)}{dx} \right) G(x) dx \quad G(x) = \int g(x) dx \quad (1.3.17)$$

$$\int x^n dx = \frac{1}{n+1} x^{n+1} + C \quad C\text{: constant given by the condition} \qquad (1.3.18)$$

$$\int \frac{1}{x} dx = \ln x + C \qquad (1.3.19)$$

$$\int e^x dx = e^x + C \qquad (1.3.20)$$

$$\int \sin(x) dx = -\cos(x) + C \quad \int \cos(x) dx = \sin(x) + C \qquad (1.3.21)$$

Using the Cartesian coordinate, the surface integral of $f(x,y)$ is given by

$$\iint f(x, y) dx dy \qquad (1.3.22)$$

Using the coordinate (p,q), eq. (1.3.22) is transformed into:

$$\iint f(x, y) dx dy = \iint g(p, q) \det \left[\begin{pmatrix} \frac{\partial x}{\partial p} & \frac{\partial x}{\partial q} \\ \frac{\partial y}{\partial p} & \frac{\partial y}{\partial q} \end{pmatrix} \right] dp dq \qquad (1.3.23)$$

where $\det[A]$ is the determinant of matrix A (see chapter 1.8).

Using the polar coordinate (r, θ) $(x = r \cos\theta, y = r \sin\theta)$, eq. (1.3.23) can be rewritten as:

$$\iint f(x, y) dx dy = \iint g(r, \theta) r dr d\theta \qquad (1.3.24)$$

When $g(r,\theta)$ has no dependence on θ, we have:

$$\iint f(x,y)dxdy = \int_0^{2\pi} d\theta \int g(r)rdr = 2\pi \int g(r)rdr \qquad (1.3.25)$$

Using the three-dimensional polar coordinates $(x = r\sin\theta\cos\phi, y = r\sin\theta\sin\phi, z = r\cos\theta)$, the volume integral is given by:

$$\iiint f(x,y,z)dxdydz = \iiint g(r,\theta,\phi)r^2\sin\theta drd\theta d\phi \qquad (1.3.26)$$

When $g(r,\theta,\phi)$ depends only on r, we have:

$$\iiint f(x,y,z)dxdydz = \int_0^{2\pi} d\phi \int_0^{\pi}\sin\theta \int g(r)r^2dr = 4\pi \int g(r)r^2dr \quad (1.3.27)$$

1.4. FIRST-ORDER DIFFERENTIAL EQUATIONS

Differential equations relate to one or more functions and their derivatives. First, we consider the first-order differential equation:

$$\frac{dy}{dx} = F(x,y) \qquad (1.4.1)$$

When $F(x,y)$ is separable function $f(x)g(y)$, eq.(1.4.1) is rewritten as:

$$\frac{dy}{dx} = f(x)g(y)$$
$$\frac{1}{g(y)}\frac{dy}{dx} = f(x)$$
$$\int \frac{1}{g(y)}\frac{dy}{dx}dx = \int \frac{1}{g(y)}dy = \int f(x)dx \qquad (1.4.2)$$

Here, we show several examples:

(1)
$$\frac{dy}{dx} = ay + b$$
$$\frac{1}{y+\frac{b}{a}}\frac{dy}{dx} = a$$

$$\int \frac{1}{y+\frac{b}{a}} dy = a \int dx$$

$$\log\left(y + \frac{b}{a}\right) = ax + C \quad y = A\left[e^{ax} - \frac{b}{a}\right] \quad A = e^C$$

(*C*: constant given by the value of *y* with a certain value of *x*) **(1.4.3)**

(2)
$$\frac{dy}{dx} = -2xy$$

$$\int \frac{1}{y} dy = -\int 2x dx$$

$$\ln(y) = -x^2 + C \quad y = e^{-x^2+C} = Ae^{-x^2}$$

for the condition $y = 1$ when $x = 0$, we have:

$$y = e^{-x^2} \tag{1.4.4}$$

(3) $\frac{dy}{dx} = \gamma y \left(1 - \frac{y}{y_\infty}\right)$ for the condition $y = y_0$ when $x = 0$, we have:

$$\int \left[\frac{1}{y\left(1-\frac{y}{y_\infty}\right)}\right] dy = \gamma \int dx$$

$$\int \left[\frac{1}{y} + \frac{1}{y_\infty - y}\right] dy = \ln(y) - \ln(y_\infty - y) = \gamma \int dx$$

$$\frac{y}{y_\infty - y} = Ae^{\gamma x}$$

$$y = \frac{y_\infty Ae^{\gamma x}}{1 + Ae^{\gamma t}} = \frac{y_\infty A}{e^{-\gamma x} + A} \quad x = 0 \quad y_0 = \frac{y_\infty A}{1+A} \quad A = \frac{y_0}{y_\infty - y_0}$$

$$y = \frac{y_0 y_\infty}{(y_\infty - y_0)e^{-\gamma x} + y_0} \tag{1.4.5}$$

These examples are special cases for which differential equations are solved using explicit formulas. Most first-order differential equations are not solvable. However, the physical equations can often be simplified (see chapter 1.6) to derive solvable equations.

1.5. SECOND-ORDER DIFFERENTIAL EQUATIONS

The physical laws are often expressed using the second-order differential equations:

$$\frac{d^2y}{dx^2} = \frac{d}{dx}\left(\frac{dy}{dx}\right) = F\left(x, y, \frac{dy}{dx}\right) \tag{1.5.1}$$

The second order homogeneous differential equation that is given as:

$$\frac{d^2y}{dx^2} + p\frac{dy}{dx} + qy = 0 \tag{1.5.2}$$

is a typical solvable second-order differential equation, that is often used for physical analysis. Note

$$\frac{de^{\lambda x}}{dx} = \lambda e^{\lambda x}, \quad \frac{d^2 e^{\lambda x}}{dx^2} = \lambda^2 e^{\lambda x}$$

$$y = c_+ e^{\lambda_+ x} + c_- e^{\lambda_- x}$$

$$\lambda_\pm^2 + p\lambda_\pm + q = 0 \quad\quad \lambda_\pm = \frac{-p \pm \sqrt{p^2 - 4q}}{2} \tag{1.5.3}$$

is the solution of eq. (1.5.2). Here we provide a solution assuming that y is the exponential function of x. The validity of this assumption is questionable. However, the uniqueness theorem [3] shows that only one solution satisfies the boundary condition (for example, values of y and dy/dx at $x = 0$). When one solution is obtained using this method, there are no other solutions.

Equation (1.5.2) can also be solved by transforming the second-order differential equation into a two-dimensional first-order differential equation:

$$\frac{dy}{dx} = z$$

$$\frac{dz}{dx} = -pz - qy \tag{1.5.4}$$

We consider the linear combination of y and z as follows:

$$\frac{d[z + \alpha y]}{dx} = (\alpha - p)z - qy = (\alpha - p)\left[z - \frac{q}{\alpha - p}y\right] \tag{1.5.5}$$

When

$$\alpha = -\frac{q}{\alpha - p} \quad \rightarrow \quad \alpha^2 - p\alpha + q = 0 \quad \alpha_\pm = \lambda_\pm + p \tag{1.5.6}$$

eq. (1.5.5) is rewritten as

$$\frac{d[z+\alpha_\pm y]}{dx} = \lambda_\pm [z + \alpha_\pm y]$$
$$z + \alpha_\pm y = C_\pm \, e^{\lambda_\pm x}$$
$$y = c_+ e^{\lambda_+ x} + c_- e^{\lambda_- x} \quad C_\pm = \pm \frac{C_\pm}{\alpha_+ - \alpha_-} \tag{1.5.7}$$

Here we assume that the real values of p, q and $p^2 - 4q < 0$. Then, eqs. (1.5.3) and (1.5.7) can be rewritten as:

$$y = e^{-\frac{px}{2}} [C_c \cos(\omega x) + C_s \sin(\omega x)] \quad \omega = \sqrt{4q - p^2} \tag{1.5.8}$$

It is generally not simple to solve the equation:

$$\frac{d^2 y}{dx^2} + p \frac{dy}{dx} + qy = f(x) \tag{1.5.9}$$

If one solution $y = g(x)$ is known, then:

$$y = c_+ e^{\lambda_+ x} + c_- e^{\lambda_- x} + g(x) \tag{1.5.10}$$

is also the solution of eq. (1.5.9). For example, we consider the equation:

$$\frac{d^2 y}{dx^2} + \omega^2 y = F \sin(\omega_0 x) \tag{1.5.11}$$

Then

$$y = \frac{F}{\omega^2 - \omega_0^2} \sin(\omega_0 x) \tag{1.5.12}$$

is one solution of eq. (1.5.11). The general solutions are given by

$$y = [C_c \cos(\omega x) + C_s \sin(\omega x)] + \frac{F}{\omega^2 - \omega_0^2} \sin(\omega_0 x) \tag{1.5.13}$$

C_c and C_s are determined from the boundary conditions (for example, values of y and dy/dx for $x = 0$).

The preceding discussions are also valid for higher-order homogeneous differential equations, which can be expressed as:

$$\sum_{k=0}^{n} a_k \frac{d^k y}{dx^k} = 0 \tag{1.5.14}$$

The solution of eq. (1.5.14) is given by:

$$y = \sum_{i=0}^{n} c_i e^{\lambda_i x}$$
$$\sum_{j=0}^{n} a_j \lambda_i^j = 0 \tag{1.5.15}$$

1.6. NUMERICAL CALCULATION OF DIFFERENTIAL EQUATIONS

For most differential equations, determining the relation between the variable x and its function y cannot be achieved using explicit formulas. However, the change in y with a slight change in x ($x \rightarrow x + h_d$, $y(x) \rightarrow y(x + h_d)$) can be determined using appropriate approximations. Repeating this procedure n times, $y(x + nh)$ leads to a numerical value. The Taylor expansion [4] is useful for evaluating the accuracy of the estimation of $y(x + h_d) - y(x)$.

The Tayler expansion gives the following relation:

$$y(x_0 + h) = y(x_0) + \left[\frac{dy}{dx}\right]_{x=x_0} h_d + \frac{1}{2}\left[\frac{d^2 y}{dx^2}\right]_{x=x_0} h_d^2 + \frac{1}{3 \cdot 2}\left[\frac{d^3 y}{dx^3}\right]_{x=x_0} h_d^3$$

$$+ \frac{1}{4 \cdot 3 \cdot 2}\left[\frac{d^4 y}{dx^4}\right]_{x=x_0} h_d^4 +$$

$$= \sum \frac{1}{n!}\left[\frac{d^n y}{dx^n}\right]_{x=x_0} h_d^n \tag{1.6.1}$$

The relation is derived from the following procedure:

$$y(x_0 + h_d) - y(x_0) = \int_{x_0}^{x_0+h_d} \frac{dy}{dx} dx \qquad (1.6.2)$$

As the first-order approximation,

$$x_0 \leq x \leq x_0 + h_d \qquad \frac{dy}{dx} = \left[\frac{dy}{dx}\right]_{x=x_0}$$
$$y(x_0 + h_d) = y(x_0) + \left[\frac{dy}{dx}\right]_{x=x_0} h_d \qquad (1.6.3)$$

is obtained. Using this method, the following simplifications are often used with $[h_d] \ll 1$:

$$(1 + h_d)^n \approx 1 + nh_d, \quad \frac{1}{1+h_d} \approx 1 - h_d, \quad \sqrt{1 + h_d} \approx 1 + \frac{h_d}{2}$$
$$e^{h_d} \approx 1 + h_d, \quad \sin(h_d) \approx h_d \qquad (1.6.4)$$

The second-order approximation is considered to improve the accuracy by:

$$x_0 \leq x \leq x_0 + h_d \qquad \frac{dy}{dx} = \left[\frac{dy}{dx}\right]_{x=x_0} + \left[\frac{d^2 y}{dx^2}\right]_{x=x_0} (x - x_0)$$

$$y(x_0 + h_d) = y(x_0) + \left[\frac{dy}{dx}\right]_{x=x_0} h_d + \frac{1}{2}\left[\frac{d^2 y}{dx^2}\right]_{x=x_0} h_d^2 \qquad (1.6.5)$$

Using the second-order Taylor's expansion, the simplification $\cos(h_d) \approx 1 - \frac{h_d^2}{2}$ is often used. By repeating this procedure, eq. (1.6.1) was obtained.

Using Euler's method, the solution of the first-order differential equation $\frac{dy}{dx} = f(x, y)$ with the condition of $y(x_0) = y_0$ is calculated as

$$x_{n+1} = x_n + h_d$$
$$y_{n+1} = y_n + f(x_n, y_n)h_d \qquad (1.6.6)$$

Using this method, only the first-order term of the Taylor expansion is considered. For example, the solution of the equation $\frac{dy}{dx} = y$ with $x_0 = 0, y_0 = 1$ (analytically solved as $y = e^x$) is calculated as:

$$x_{n+1} = x_n + h_d$$

$$y_{n+1} = y_n + y_n h_d = y_n(1 + h_d) \tag{1.6.7}$$

Then we have,

$$x_n = n h_d, \quad y_n = (1 + h_d)^n = \left(1 + \frac{x_n}{n}\right)^n \tag{1.6.8}$$

For $h_d \to 0 (n \to \infty)$, eq. (1.6.8) converges to $y_n = e^{x_n}$ (see eq. (1.3.7)), which corresponds to the analytical solution.

Euler's method is simple, but it is not sufficiently accurate for practical application. The accuracy can be improved using the middle-point method, which is given by:

$$\begin{aligned} x_{n+1} &= x_n + h_d \\ y_{n+1} &= y_n + f\left(x_n + \frac{h_d}{2}, y_n + \frac{h_d}{2} f(x_n, y_n)\right) h_d \end{aligned} \tag{1.6.9}$$

As shown in the following, eq. (1.6.9) is accurate for the first-and second-order terms of the Taylor expansion. By performing the first-order Taylor expansion of $f(x,y)$, we have:

$$\begin{aligned} f\left(x_n + \frac{h_d}{2}, y_n + \frac{h_d}{2} f(x_n, y_n)\right) &= f(x_n, y_n) + \frac{\partial f(x_n, y_n)}{\partial x} \frac{h_d}{2} + \\ \frac{\partial f(x_n, y_n)}{\partial y} f(x_n, y_n) \frac{h_d}{2} \\ \left(\text{considering } f(x_n, y_n) = \frac{dy_n}{dx_n}\right) \\ &= f(x_n, y_n) + \frac{\partial f(x_n, y_n)}{\partial x} \frac{h_d}{2} + \frac{\partial f(x_n, y_n)}{\partial y} \frac{dy_n}{dx_n} \frac{h_d}{2} \\ &= f(x_n, y_n) + \frac{df(x_n, y_n)}{dx} \frac{h_d}{2} \end{aligned} \tag{1.6.10}$$

Using eq. (1.6.10), eq. (1.6.9) is rewritten as:

$$y_{n+1} = y_n + f(x_n, y_n) h_d + \frac{1}{2} \frac{df(x_n, y_n)}{dx} h_d^2 = y_n + \frac{dy_n}{dx_n} h_d + \frac{1}{2} \frac{d^2 y_n}{dx_n} h_d^2 \tag{1.6.11}$$

The solution of the equation $\frac{dy}{dx} = y$ with $x_0 = 0, y_0 = 1$ is calculated as:

$$y_{n+1} = y_n + \left(y_n + \frac{1}{2} y_n h_d\right) h_d = y_n \left(1 + h_d + \frac{h_d^2}{2}\right) \tag{1.6.12}$$

The accuracy obtained when using the middle point method is much higher than that obtained with Euler's method, for which only the first-order term is considered.

The Runge-Kutta method has been developed, wherein y_{n+1} - y_n is estimated with higher accuracy [5]:

$$y_{n+1} = y_n + h_d \sum_{i=1}^{s} b_i k_i$$
$$k_1 = f(x_n, y_n)$$
$$k_2 = f(x_n + c_2 h_d, y_n + h_d(a_{21} k_1))$$
$$k_3 = f(x_n + c_3 h_d, y_n + h_d(a_{31} k_1 + a_{32} k_2))$$

$$\vdots$$

$$k_s = f(x_n + c_s h_d, y_n + h_d(a_{s1} k_1 + a_{s2} k_2 + \cdots a_{ss-1} k_{s-1})) \quad \text{(1.6.13)}$$

The fourth-order Runge-Kutta method is as follows [5]:

$$y_{n+1} = y_n + \frac{h_d}{6}(k_1 + 2k_2 + 2k_3 + k_4)$$
$$k_1 = f(x_n, y_n)$$
$$k_2 = f\left(x_n + \frac{1}{2}h_d, y_n + \frac{h_d}{2}k_1\right)$$
$$k_3 = f\left(x_n + \frac{1}{2}h_d, y_n + \frac{h_d}{2}k_2\right)$$
$$k_4 = f(x_n + h_d, y_n + h_d k_3) \quad \text{(1.6.14)}$$

is often used because y_{n+1} - y_n is estimated corresponding to the first to the fourth-order term of the Tayler expansion. The solution of the equation $\frac{dy}{dx} = y$ is calculated as:

$$y_{n+1} = y_n \left(1 + h_d + \frac{h_d^2}{2} + \frac{h_d^3}{6} + \frac{h_d^4}{24}\right) \quad \text{(1.6.15)}$$

Fig. (**1.3**). shows the error of the numerical solution of the equation $\frac{dy}{dx} = y$ with $x_0 = 0, y_0 = 1$ (analytical solution $y = e^x$) at $0 < x < 2$ using Euler's method, the middle point method, and the fourth-order Runge-Kutta method with $h = 0.01$. Using the middle point method (fourth-order Runge-Kutta method), the calculation is performed with an error two (eight) orders smaller compared to Euler's method.

Error from the analytical solution

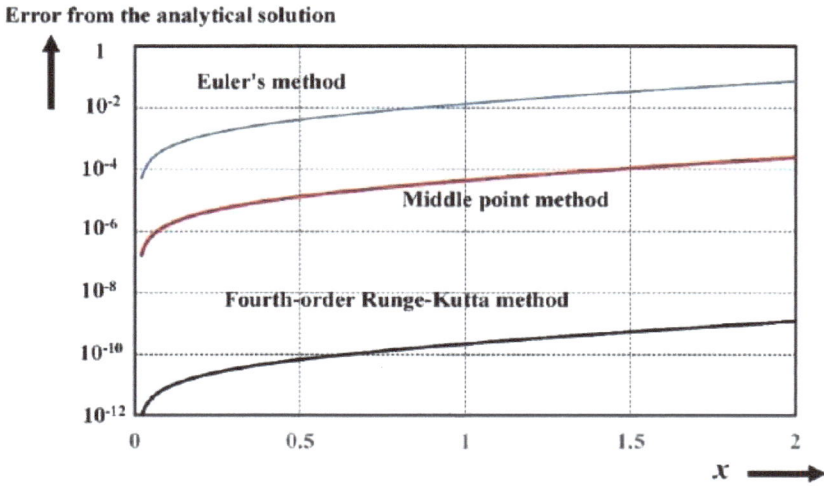

Fig. (1.3). The errors of numerical calculations based on the Euler's method, the middle point method, and the fourth-order Runge-Kutta method with $h = 0.01$.

The accuracy is improved when higher-order Runge-Kutta methods are used, but the treatments are much more complicated compared to the fourth-order Runge-Kutta method.

In principle, the error is reduced by taking a small value of h_d, which increases the calculation cycle number. A large number of calculation cycles increases the calculation time. It is questionable if the smaller value of h_d always reduces the error because the error of the numerical calculation is increased by repeating the calculation loop. For numerical calculations, there is also an error called the "rounding error", which is caused by the approximation of irrational numbers. For example, $(\sqrt{2})^2 = 2$ is not always obtained when computer calculation is performed. Therefore, some ingenuity is required for numerical calculations. To confirm the reliability of the results, it is useful to evaluate the consistency of the process by examining the solution for different values of h_d.

The higher-order differential equations should be transformed into multi-dimensional first-order differential equations. For example, the equation

$$\frac{d^2y}{dx^2} + \sin(y) = 0 \qquad (1.6.16)$$

can be transformed to

$$\frac{dy}{dx} = z, \quad \frac{dz}{dx} = -\sin(y) \tag{1.6.17}$$

Using Euler's method, the solution of this equation is calculated as:

$$y_{n+1} = y_n + z_n h_d, \quad z_{n+1} = z_n - \sin(y_n)\, h_d \tag{1.6.18}$$

When $|y| \ll 1$, eq. (1.6.16) is approximated as:

$$\frac{d^2 y}{dx^2} + y = 0 \tag{1.6.19}$$

For the condition $x = 0, y = y_0 (\ll 1), \frac{dy}{dx} = 0$, the approximated solution is given by:

$$y = y_0 \cos(x) \tag{1.6.20}$$

1.7. CHAOS

The differential equations that are used to describe future phenomena can be solved numerically. However, there is a case wherein future predictions are physically impossible, called "chaos." This phenomenon is caused by the nonlinearity of the equation; a slight difference in the initial value results in an exponential growth of the difference with the evolution of time.

The concept of chaos was introduced using Lorenz's equation to determine the weather in the future [6]. Weather is predictable for the next week, but predictions on a longer time scale are not reliable. This unpredictability was expressed by the statement, "Does the Flap of a Butterfly's Wings in Brazil set off a Tornado in Texas?"

As the simplest example of chaos, we show a repetition of the baker's transformation as follows (see Fig. **1.4**) [7]:

(1) We consider the position of x_n on a bar, with both ends at 0 and 1.
(2) The bar is expanded to double the length, then collapsed to half. Then x_n is transformed to $x_{n+1} = 1 - |1 - 2x_n|$.

Fig. (**1.5**). shows x_n for $x_0 = 0.3$ and 0.300003. With $n > 15$, the difference of x_n becomes significant. Given that x_0 must have some uncertainty, the value of x_n

becomes unpredictable after this transformation is repeated. The concept of the baker's transformation shows that the chaos phenomenon is observed as a function of time $x(t,x_0)$ (x_0: initial value of x), satisfying $x_{mim} < x(t,x_0) < x_{max}$ and $|x(t,x_0 + \delta x) - x(t,x_0)| = e^{(t/t_0)} \delta x$ ($t_0 > 0$). The value of t_0 (= $\ln[|x_{max} - x_{min}|/\delta x]$) is called the Lyapunov exponent, which is a parameter such that we can predict the value of $x(t)$ only for $t < t_0$.

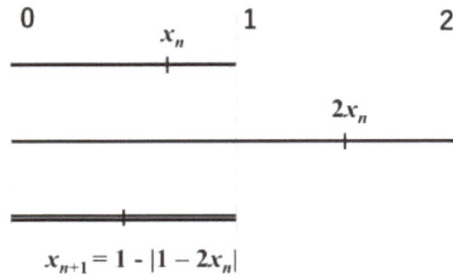

Fig. (1.4). Baker's transformation. (This figure is used also in "Measurement, Uncertainty and Lasers" by M. Kajita).

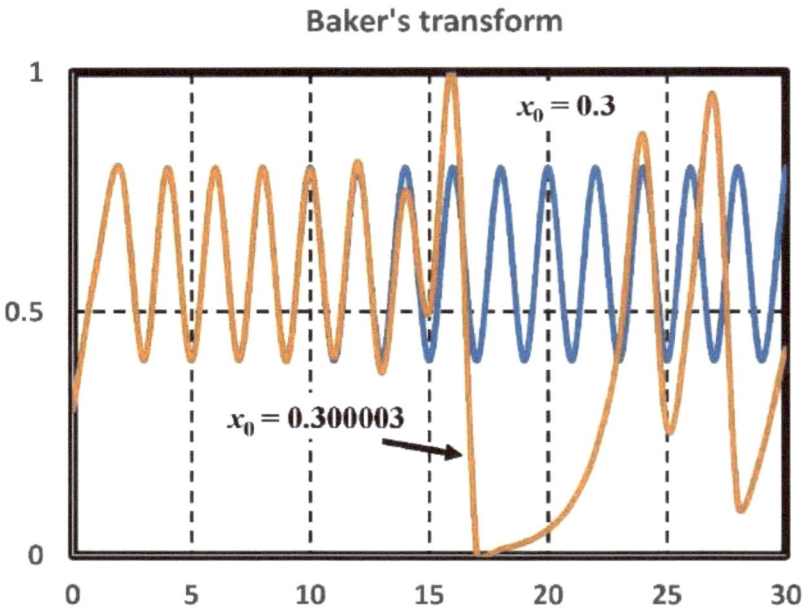

Fig. (1.5). Baker's transformation with the initial values of 0.3 and 0.300003. (This figure is used also in "Measurement, Uncertainty and Lasers" by M. Kajita).

Chaotic solutions are obtained using n-th order ordinary differential equations ($n > 3$) with a single scalar variable [8]. For example, a solution to

$$\frac{d^4y}{dx^4} + \frac{d^3y}{dx^3} + A\frac{d^2y}{dx^2} + \left(\frac{dy}{dx}\right)^2 + x = 0 \tag{1.7.1}$$

is chaotic with $3.3 < A < 4.6$ is chaotic.

1.8. EIGENVALUES AND EIGENVECTORS OF MATRICES

Matrices are used to describe the linear combinations of parameters. For example, the linear combination of two parameters (x_1, x_2) is described using a 2×2 matrix A and a two-dimensional vector $\vec{x}$ as follows:

$$A = \begin{pmatrix} a_{11} & a_{12} \\ a_{21} & a_{22} \end{pmatrix} \quad \vec{x} = \begin{pmatrix} x_1 \\ x_2 \end{pmatrix} \quad A\vec{x} = \begin{pmatrix} a_{11}x_1 + a_{12}x_2 \\ a_{21}x_1 + a_{22}x_2 \end{pmatrix} \tag{1.8.1}$$

Between the matrices A and B,

$$A = \begin{pmatrix} a_{11} & a_{12} \\ a_{21} & a_{22} \end{pmatrix} \quad B = \begin{pmatrix} b_{11} & b_{12} \\ b_{21} & b_{22} \end{pmatrix} \quad A + B = \begin{pmatrix} a_{11} + b_{11} & a_{12} + b_{12} \\ a_{21} + b_{21} & a_{22} + b_{22} \end{pmatrix}$$
$$AB = \begin{pmatrix} a_{11}b_{11} + a_{12}b_{21} & a_{11}b_{12} + a_{12}b_{22} \\ a_{21}b_{11} + a_{22}b_{21} & a_{21}b_{12} + a_{22}b_{22} \end{pmatrix} \tag{1.8.2}$$

Generally, $AB \neq BA$, and the Dirac equation (see chapter 5) was derived using matrices that satisfy $AB + BA = 0$.

From eq. (1.8.1),

$$I = \begin{pmatrix} 1 & 0 \\ 0 & 1 \end{pmatrix} \quad I\vec{x} = \vec{x} \quad IA = AI = A \tag{1.8.3}$$

and the inverse matrix of A (A^{-1}) is defined from:

$$A^{-1}A = AA^{-1} = I \tag{1.8.4}$$

For the 2×2 matrix,

$$A^{-1} = \frac{1}{\det(A)} \begin{pmatrix} a_{22} & -a_{12} \\ -a_{21} & a_{11} \end{pmatrix}$$
$$\det(A) = a_{11}a_{22} - a_{12}a_{21} \tag{1.8.5}$$

where det(A) is called "determinant", which is the product of eigenvalues, that is introduced below.

$$\det(AB) = \det(A)\det(B) \tag{1.8.6}$$

is always satisfied. When $\det(A) = 0$, A^{-1} does not exist.

Here we introduce the eigenvalues and eigenvectors of the matrix. For a certain combination of values and vectors,

$$A\overrightarrow{x_e} = \lambda_e \overrightarrow{x_e} \tag{1.8.7}$$

holds. Here, λ_e is the eigenvalue and $\overrightarrow{x_e}$ is the eigenvector. Equation (1.8.7) can be rewritten as:

$$\begin{aligned}
[A - \lambda_e I]\overrightarrow{x_e} = \vec{0} \quad \vec{0} = \begin{pmatrix} 0 \\ 0 \end{pmatrix} \\
(a_{11} - \lambda_e)x_{e1} + a_{12}x_{e2} = 0 \\
a_{21}x_{e1} + (a_{22} - \lambda_e)x_{e2} = 0
\end{aligned} \tag{1.8.8}$$

To satisfy eq. (1.8.8) with $\overrightarrow{x_e} \neq 0$, the eigenvalue λ_e is obtained as follows:

$$\begin{aligned}
(a_{11} - \lambda_e){:}\, a_{12} = a_{21}{:}\,(a_{22} - \lambda_e) \\
(\lambda_e - a_{11})(\lambda_e - a_{22}) - a_{12}a_{21} = 0
\end{aligned} \tag{1.8.9}$$

Equation (1.8.9) is equivalent to:

$$\det(A - \lambda_e I) = 0 \tag{1.8.10}$$

which is valid with any $N \times N$ matrix. Otherwise, $(A - \lambda_e I)^{-1}$ must exist and $\vec{x}$ is obtained by $(A - \lambda_e I)^{-1}\vec{0}$, which is not possible except when $\vec{x} = \vec{0}$.

The ratio between x_{e1} and x_{e2} for the corresponding eigenvalue $\lambda_{e1,2}$ is given by eq. (1.8.9).

EXERCISE

(1) Numerical calculations of

$$\frac{dy}{dx} = -2xy$$

can be performed using $x_{n+1} = x_n + h$.

Determine the formula for obtaining y_{n+1} using y_n, x_n, and h when Euler's method and the midpoint method are used.

(Answer)

Euler's method

$$y_{n+1} = y_n - 2x_n y_n h_d$$

middle-point method

$$y_{n+1} = y_n - 2\left(x_n + \frac{h_d}{2}\right)(y_n - x_n y_n h_d) h_d$$

$$= y_n - 2x_n y_n h_d + (4x_n^2 y_n - 2y_n)\frac{h_d^2}{2} + x_n y_n h_d^3$$

(2) Obtain the eigenvalue of the matrix

$$\begin{pmatrix} \Delta & \Omega \\ \Omega & 0 \end{pmatrix}$$

The eigenvalue of this matrix has a very important role in quantum mechanics.

(Answer)

$$\frac{\Delta \pm \sqrt{\Delta^2 + \Omega^2}}{2}$$

REFERENCES

[1]   J.B. Seaborn, "Mathematics for physical science", *Smostly Physisists pringer,* 2002.
       [http://dx.doi.org/10.1007/978-1-4684-9279-8]

[2]   Y. Zeldovich, and I. Yaglom, "Higher math for beginners: mostly physisists and engineers", MIR, ASIN: B07116W6T2

[3]   E. V. Weisstein, "Uniqueness theorem. mathworld, wolfram.com", Retrieved 2019-11-29.

[4]   E. Abdi, *Linear Algebra for Neural Networks.* International Encyclopedia of Social & Behavoral Science, 2001.
       [http://dx.doi.org/10.1016/B0-08-043076-7/00609-4]

[5]   E. Sueli, and D. Mayers, "An introduction to numerical analysis", *Cambridge University Press.* ISBN 0-521-00794-1.

[6]   E.N. Lorenz, "Deterministic nonperiodic flow", *J. Atmos. Sci.,* vol. 20, no. 2, pp. 130-41.1, 1963.
       [http://dx.doi.org/10.1175/1520-0469(1963)020<0130:DNF>2.0.CO;2]

[7]   G. Radons, G.C. Hartmann, H.H. Diebner, and O.E. Rossler, "Staircase baker's map generates flaring-type time series", *Discrete Dyn. Nat. Soc.,* vol. 5, no. 2, pp. 107-120, 2000.
       [http://dx.doi.org/10.1155/S1026022600000467]

[8]   K.E. Chlouverakis, and J.C. Sprott, "Chaotic hyperjerk systems", *Chaos Solitons Fractals,* vol. 28, no. 3, pp. 739-746, 2006. http://sprott.physics.wisc.edu/pubs/paper297.htm
       [http://dx.doi.org/10.1016/j.chaos.2005.08.019]

CHAPTER 2

Analysis in the Classical Mechanics

Abstract: Equations of motion with arbitrary potential fields are described by simple formulas based on Newtonian law. However, they are generally complicated to solve. This chapter presents several methods for deriving and solving equations of motion with different coordinate systems.

Two-body motion is introduced based on the coordinates of the motion of the center of mass and the relative motion. R elative motion perpendicular to a relative position is discussed using angular momentum, wherein the temporal change is given by the torque. The Lagrange equation is introduced for application to all types of coordinates. The polar coordinate system is convenient for deriving an equation of motion with a spherically symmetric potential because the angular momentum is constant. The motion in the radial direction should be considered taking the centrifugal force potential into account. The Lagrangian is also introduced with respect to electromagnetic fields.

Keywords: Angular momentum, Center of mass, Centrifugal force, Energy, Gravity, Kepler's law, Lagrange equation, Momentum. Inertial moment, Newtonian mechanics, Relative motion.

2.1. FUNDAMENTAL OF CLASSICAL MECHANICS

Newton derived three laws of motion in his monumental work Philosophiæ Naturalis Principia Mathematica (Mathematical Principles of Natural Philosophy), which was published in 1687 [1].

1. An object either remains at rest or continues to move in a straight line at a constant velocity unless acted on by a net external force.

2. The sum of the forces $\vec{F}$ acting on an object is equal to the mass m of that object multiplied by the acceleration $\vec{a}$ of the object,

3. When a body exerts a force on a second body, the second body simultaneously exerts a force equal in magnitude and opposite in direction on the first body.

Using the Cartesian coordinate $\vec{r} = (x, y, z)$, the second law is given by the equation

Masatoshi Kajita
All rights reserved-© 2022 Bentham Science Publishers

$$m\frac{d\vec{v}}{dt} = m\frac{d^2\vec{r}}{dt^2} = \vec{F} \quad \vec{v} = \frac{d\vec{r}}{dt} = \begin{pmatrix} v_x \\ v_y \\ v_z \end{pmatrix} \quad \vec{F} = \begin{pmatrix} F_x \\ F_y \\ F_z \end{pmatrix}$$

(2.1.1)

where m is the mass and t is the time. The first law corresponds to the second law with $\vec{F} = 0$, although it was initially given as the definition of the "inertial frame". In classical mechanics, momentum is defined as follows (this definition is not correct in the case of relativistic theory and quantum mechanics):

$$\vec{p} = m\vec{v}$$

(2.1.2)

From the force, the potential energy P_E is defined as:

$$P_E = -\int F_x dx - \int F_y dy - \int F_z dz$$

(2.1.3)

Then eq. (2.1.1) can be rewritten as:

$$\frac{d\vec{p}}{dt} = -\nabla P_E \quad \nabla = \begin{pmatrix} \frac{\partial}{\partial x} \\ \frac{\partial}{\partial y} \\ \frac{\partial}{\partial z} \end{pmatrix}$$

(2.1.4)

The temporal change of $|\vec{p}|^2$ by a force is given by:

$$\frac{d|p|^2}{dt} = 2p_x \frac{dp_x}{dt} + 2p_y \frac{dp_y}{dt} + 2p_z \frac{dp_z}{dt}$$

$$\text{using} \quad \frac{dp_q}{dt} = -\frac{\partial P_E}{\partial q}, p_q = m\frac{dq}{dt}$$

$$= -2m\frac{\partial P_E}{\partial x}\frac{dx}{dt} - 2m\frac{\partial P_E}{\partial y}\frac{dy}{dt} - 2m\frac{\partial P_E}{\partial z}\frac{dz}{dt}$$

(2.1.5)

Defining the kinetic energy as follows:

$$K_E = \frac{|\vec{p}|^2}{2m} = \frac{m|\vec{v}|^2}{2}$$

(2.1.6)

the temporal change of the potential energy and the total energy $E = K_E + P_E$ are given by:

$$\frac{dP_E}{dt} = \frac{\partial P_E}{\partial t} + \left[\frac{\partial P_E}{\partial x}\frac{dx}{dt} + \frac{\partial P_E}{\partial y}\frac{dy}{dt} + \frac{\partial P_E}{\partial z}\frac{dz}{dt}\right] = \frac{\partial P_E}{\partial t} - \frac{dK_E}{dt} \qquad \text{(see eq. (1.3.10))}$$

$$\frac{dE}{dt} = \frac{dK_E}{dt} + \frac{dP_E}{dt} = \frac{\partial P_E}{\partial t} \qquad\qquad \textbf{(2.1.7)}$$

The total energy is conserved when the P_E has no dependence on time (constant potential energy distribution). Considering the interaction between two bodies a and b, the sum of their momenta is conserved because the change in momentum is given by:

$$\Delta\overrightarrow{p_a} = \int \overrightarrow{F_{b\to a}}dt \qquad \Delta\overrightarrow{p_b} = \int \overrightarrow{F_{a\to b}}dt$$

$$\overrightarrow{F_{a\to b}} = -\overrightarrow{F_{b\to a}} \quad \text{(from the Newtonian third law)} \qquad \textbf{(2.1.8)}$$

According to the theory of relativity and quantum mechanics, the correspondence between the three-dimensional position vector $\vec{r}$ and the momentum vector $\vec{p}$ can be expanded to the relation between the four-dimensional position vectors $\overrightarrow{r^4} = (x, y, z, ct)$ and $\overrightarrow{p^4} = (p_x, p_y, p_z, E/c)$, where c is the speed of light in a vacuum. For two interacting bodies, the sum of the four-dimensional momentum vector must be conserved. What is the correspondence between time and energy? Equation (2.1.4) shows that the change in $\vec{p}$ is given by the dependence of P_E on $\vec{r}$. However, E changes when there is a temporal change in P_E. We can say that the components of $\overrightarrow{p^4}$ change when the potential energy changes due to the corresponding components of $\overrightarrow{r^4}$.

2.2. TWO-BODY MOTION AND TORQUE

Here, we consider two-body motion (masses: m_1 and m_2) based on the motion of the center of mass, binding, and rotation. Bodies 1 and 2 experience the external forces $\overrightarrow{F_1}\left(= \overrightarrow{\delta F_1} + \overrightarrow{F_{int}}\right)$ and $\overrightarrow{F_2}\left(= \overrightarrow{\delta F_2} - \overrightarrow{F_{int}}\right)$, respectively, as shown in Fig. (2.1). Here, $\pm\overrightarrow{F_{int}}$ is the interaction force between the two bodies.

$$m_1\frac{d^2\overrightarrow{r_1}}{dt^2} = \overrightarrow{F_1} \qquad\qquad \textbf{(2.2.1)}$$

$$m_2 \frac{d^2 \vec{r_2}}{dt^2} = \vec{F_2} \tag{2.2.2}$$

$(2.2.1) + (2.2.2)$

$$(m_1 + m_2)\frac{d^2 \vec{R}}{dt^2} = \vec{F_1} + \vec{F_2} = \vec{\delta F_1} + \vec{\delta F_2}$$

$$\vec{R} = \frac{m_1 \vec{r_1} + m_2 \vec{r_2}}{m_1 + m_2} \tag{2.2.3}$$

$$(2.2.1)\times \frac{m_2}{m_1 + m_2} - (2.2.2) \times \frac{m_1}{m_1 + m_2}$$

$$\mu_r \frac{d^2 \vec{r}}{dt^2} = \frac{m_2}{m_1 + m_2}\vec{F_1} - \frac{m_1}{m_1 + m_2}\vec{F_2} = \vec{F_{int}} + \frac{m_2}{m_1 + m_2}\vec{\delta F_1} - \frac{m_1}{m_1 + m_2}\vec{\delta F_2}$$

$$\vec{r} = \vec{r_1} - \vec{r_2} \quad \mu_r = \frac{m_1 m_2}{m_1 + m_2} \tag{2.2.4}$$

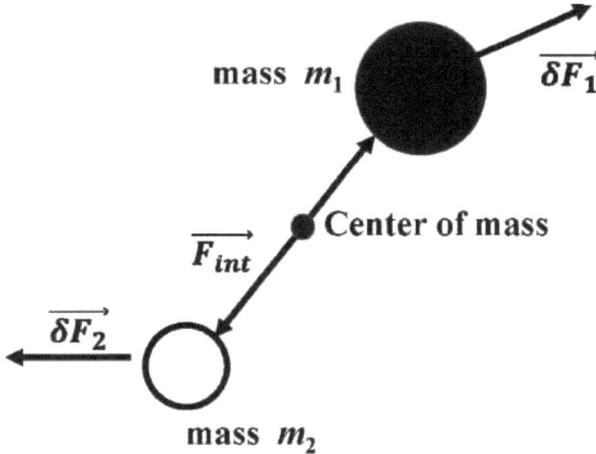

Fig. (2.1). Diagram showing the forces acting on the to two bodies.

Here, $\vec{R}$ denotes the position of the center of mass and $\vec{r}$ is the relative position. μ_r is the reduced mass. Equation (2.2.3) shows the motion of the center of mass, considering the binding of the two bodies to produce a single body. Equation (2.2.4) shows the relative motion between two bodies. The direction of $\vec{F_{int}}$ is parallel to $\vec{r}$. As the simplest case, we can consider the harmonic force that provides the minimum potential at $\vec{r} = \vec{r_0}$;

$$\overrightarrow{F_{int}} = -k(\vec{r} - \vec{r_0}) \quad (k: \text{spring constant}) \tag{2.2.5}$$

For a harmonic force, the temporal change in $|\vec{r}|$ is given by a sinusoidal function. The direction of the force, which is represented as:

$$\overrightarrow{F_r} = \frac{m_2}{m_1+m_2} \overrightarrow{\delta F_1} - \frac{m_1}{m_1+m_2} \overrightarrow{\delta F_2} \tag{2.2.6}$$

is perpendicular to the direction of the vector $\vec{r}$, so that Newton's third law holds. $\overrightarrow{F_r}$ is the force that changes the rotational angular velocity. No change in the rotational angular velocity was induced when $\overrightarrow{\delta F_2} = \frac{m_2}{m_1} \overrightarrow{\delta F_1}$. This change is often discussed using the angular momentum on the surface perpendicular to the vector's direction.

$$\vec{L} = I_c \frac{d\vec{\theta}}{dt} = \vec{r} \times \overrightarrow{p_r} = \vec{r} \times \mu_r \frac{d\vec{r}}{dt}$$

(p_r: momentum of the relative motion, $\vec{\theta}$: angular vector that represents the direction of $\vec{r}$ on the rotating surface) $\tag{2.2.7}$

where I_c is the inertial moment around the center of mass given by

$$I_c = \mu_r [\vec{r}]^2 = m_1 \left| \overrightarrow{r_1} - \vec{R} \right|^2 + m_2 \left| \overrightarrow{r_2} - \vec{R} \right|^2 \tag{2.2.8}$$

The temporal change of the angular momentum is given by

$$\frac{d\vec{L}}{dt} = \frac{d\vec{r}}{dt} \times \overrightarrow{p_r} + \vec{r} \times \frac{d\overrightarrow{p_r}}{dt} = \overrightarrow{N_r} \quad \left(\frac{d\vec{r}}{dt} \times \overrightarrow{p_r} = 0 \right) \tag{2.2.9}$$

where $\overrightarrow{N_r}$ is the torque around the center of mass, which is given by:

$$\overrightarrow{N_r} = \vec{r} \times \overrightarrow{F_r} = (\overrightarrow{r_1} - \vec{R}) \times \overrightarrow{\delta F_1} - (\overrightarrow{r_2} - \vec{R}) \times \overrightarrow{\delta F_2} \tag{2.2.10}$$

which is derived from eq. (2.2.6). When the potential energy has no dependence on θ, $\overrightarrow{N_r} = 0$, and the angular momentum is constant (see chapter 2.3).

By transforming the coordinate, the motion of the two-body system can be separated into the motion of the center of mass, the vibrational motion induced by the bonding force, and the rotational motion induced by the torque.

The inertial momentum and torque were defined around the center of mass using eqs. (2.2.7) and (2.2.9). Both parameters around point $\overrightarrow{R_f} = \vec{R} + \overrightarrow{\delta R}$ are given as follows:

$$I_{cf} = m_1\left[\overrightarrow{r_1} - \overrightarrow{R_f}\right]^2 + m_2\left[\overrightarrow{r_2} - \overrightarrow{R_f}\right]^2 = I_c + (m_1 + m_2)\left|\overrightarrow{\delta R}\right|^2$$
$$\overrightarrow{N_{rf}} = \left(\overrightarrow{r_1} - \overrightarrow{R_f}\right) \times \overrightarrow{\delta F_1} - \left(\overrightarrow{r_2} - \overrightarrow{R_f}\right) \times \overrightarrow{\delta F_2} \tag{2.2.11}$$

2.3. LAGRANGE EQUATION

In Newtonian mechanics, the equation of motion using Cartesian coordinates is given by eq. (2.1.1). However, it is often convenient to describe the potential energy using other coordinates. For example, polar coordinates are more convenient than Cartesian coordinates when the potential is spherically symmetric (*i.e.*, depends only on distance). The Lagrange equation is useful for deriving the equations for all types of coordinate systems. The Lagrangian is defined as:

$$L_g = K_E - P_E \tag{2.3.1}$$

and the Lagrange equation is given by

$$\frac{d}{dt}\left(\frac{\partial L_g}{\partial \dot{q}}\right) - \frac{\partial L_g}{\partial q} = 0 \quad \dot{q} = \frac{dq}{dt} \tag{2.3.2}$$

where q is the position parameter for any coordinate system. Using Cartesian coordinate,

$$L_g = \frac{m}{2}\dot{x}^2 - P_E(x) \tag{2.3.3}$$

and

$$\frac{d}{dt}\left(\frac{\partial L_g}{\partial \dot{x}}\right) - \frac{\partial L_g}{\partial x} = m\frac{d^2 x}{dt^2} + \frac{\partial P_E}{\partial x} = 0 \qquad m\frac{d^2 x}{dt^2} = -\frac{\partial P_E}{\partial x} = F_x \qquad (2.3.4)$$

which corresponds to eq. (2.1.1). When a body can move only in an orbit $(x, y(x))$ with gravitational potential $(P_E = mgy)$, the equation of motion is derived as follows:

$$L_g = \frac{m}{2}\left[1 + \left(\frac{dy}{dx}\right)^2\right]\dot{x}^2 - mgy$$

$$\frac{d}{dt}\left(\frac{\partial L_g}{\partial \dot{x}}\right) - \frac{\partial L_g}{\partial x} = m\left[1 + \left(\frac{dy}{dx}\right)^2\right]\frac{d^2 x}{dt^2} + mg\frac{dy}{dx} = 0 \qquad (2.3.5)$$

The advantage of the Lagrange equation is that it applies to all types of coordinate systems. Using an appropriate coordinate s, the Lagrange equation can be transformed to a combination of coordinates using (x, y, z) as follows, and its validity can be confirmed mathematically:

$$\frac{d}{dt}\left(\frac{\partial L_g}{\partial \dot{s}}\right) - \frac{\partial L_g}{\partial s} = \frac{d}{dt}\left[\frac{\partial L_g}{\partial \dot{x}}\frac{\partial \dot{x}}{\partial \dot{s}} + \frac{\partial L_g}{\partial \dot{y}}\frac{\partial \dot{y}}{\partial \dot{s}} + \frac{\partial L_g}{\partial \dot{z}}\frac{\partial \dot{z}}{\partial \dot{s}}\right] - \left(\frac{\partial L_g}{\partial x}\frac{\partial x}{\partial s} + \frac{\partial L_g}{\partial y}\frac{\partial y}{\partial s} + \frac{\partial L_g}{\partial z}\frac{\partial z}{\partial s}\right)$$

using $\frac{\partial \dot{q}}{\partial \dot{s}} = \frac{\partial q}{\partial s}$ $(q = x, y, z)$

and $\frac{d}{dt}\left(\frac{\partial L_g}{\partial \dot{q}}\right) - \frac{\partial L_g}{\partial q} = 0$

$$= \left(\frac{\partial x}{\partial s}\right)\left[\frac{d}{dt}\left(\frac{\partial L_g}{\partial \dot{x}}\right) - \frac{\partial L_g}{\partial x}\right] + \left(\frac{\partial y}{\partial s}\right)\left[\frac{d}{dt}\left(\frac{\partial L_g}{\partial \dot{y}}\right) - \frac{\partial L_g}{\partial y}\right] + \left(\frac{\partial z}{\partial s}\right)\left[\frac{d}{dt}\left(\frac{\partial L_g}{\partial \dot{z}}\right) - \frac{\partial L_g}{\partial z}\right] = 0 \qquad (2.3.6)$$

Here we consider the two-dimensional polar coordinate (r, θ) $(x = r\cos\theta, y = r\sin\theta)$. Thus, we have:

$$K_E = \frac{m}{2}\left[\dot{r}^2 + r^2\dot{\theta}^2\right]$$
$$L_g = \frac{m}{2}\left[\dot{r}^2 + r^2\dot{\theta}^2\right] - P_E(r, \theta) \qquad (2.3.7)$$

When P_E has no dependence on θ, we have:

$$m\frac{d^2 r}{dt^2} - mr\left(\frac{d\theta}{dt}\right)^2 + \frac{dP_E(r)}{dr} = 0 \qquad (2.3.8)$$

$$\frac{dL}{dt} = 0 \qquad L = \left[mr^2\left(\frac{d\theta}{dt}\right)\right] \qquad (2.3.9)$$

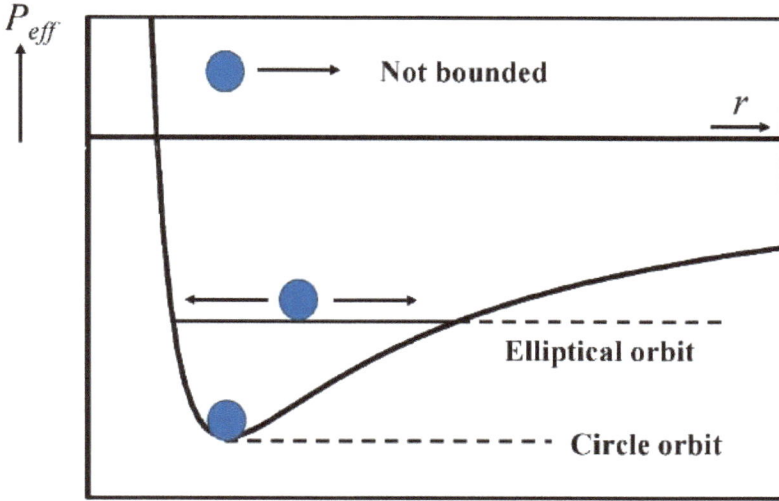

Fig. (2.2). The effective potential P_{eff} as a function of r assuming that the potential P_E is proportional to $1/r$. The orbit of the body depends on the total energy $E_{tot.}$ When $E_{tot} > 0$, it is not bounded. The orbit is circular when $E_{tot} = \min_r P_{eff}$. When $\min_r P_{eff} < E_{tot} < 0$ there are two solutions of r for $K_r = 0$ and the orbit is elliptical.

The second term of eq. (2.3.8) represents a centrifugal force. Equation (2.3.9) shows the constancy of the angular momentum L (see chapter 2.2). Using L, eq. (2.3.8) can be rewritten as:

$$m\frac{d^2r}{dt^2} = \frac{L^2}{mr^3} - \frac{dP_E(r)}{dr} = -\frac{d}{dr}P_{eff}(r)$$
$$P_{eff}(r) = \frac{L^2}{2mr^2} + P_E(r) \tag{2.3.10}$$

where $P_{eff}(r)$ is the effective potential considering the centrifugal potential into account, as shown in Fig. (2.2). Here, we consider the orbit with different total energies E_{tot} assuming

$$P_E(r) = -\frac{C}{r^k} \ (C > 0, k > 0),$$
$$E_{tot} = K_r + P_{eff} \geq P_{eff} \quad K_r = \frac{m}{2}\left(\frac{dr}{dt}\right)^2 \tag{2.3.11}$$

As shown in Fig. (**2.2**) the orbit with different values of E_{tot} is given by:

$$E_{tot} = \min_{r} P_{eff} \qquad K_r = 0 \qquad \text{circular orbit}$$
$$\min_{r} P_{eff} < E_{tot} < 0 \quad \text{elliptical orbit}$$
$$E_{tot} \geq 0 \qquad \text{non-bonding.}$$

Kepler published three laws related to planetary motion.

1. The orbit of a planet is an ellipse with the sun at one of the two foci.
2. A line segment joins a planet, and the sun sweeps out equal areas during equal time intervals.
3. The square of the orbital period T of a planet is proportional to the cube of the semi-major axis of the orbit.

The first law is derived from eq. (2.3.10) with $\min_{r} P_{eff} < E_{tot} < 0$ and the second law denotes the constancy of the angular momentum (see eq. (2.3.9)). From the third law, the gravitational potential energy is derived from being proportional to $1/r$, as follows. If we consider circular orbit assuming $P_E \propto \frac{1}{r^k}$ (r: radius of the orbit). Then, the relation between T and r is given by:

$$mr\left(\frac{d\theta}{dt}\right)^2 = \frac{dP_E(r)}{dr}$$
$$\text{using} \quad \frac{d\theta}{dt} = \frac{2\pi}{T} \quad \frac{dP_E(r)}{dr} = (k+1)\frac{C}{r^{k+1}}$$
$$mr\left(\frac{2\pi}{T}\right)^2 = (k+1)\frac{C}{r^{k+1}}$$
$$r^{2+k} = \frac{(k+1)C}{m}\left(\frac{T}{2\pi}\right)^2 \qquad (2.3.12)$$

By comparing Kepler's third law and eq. (2.3.12), $P_E(r) \propto \frac{1}{r}$ ($k = 1$) is derived. The detailed derivation of Kepler's law is presented in Ref. [2].

2.4.   APPLICATION   OF   LAGRANGE   EQUATION   TO ELECTROMAGNETIC FIELDS

A moving mass with a velocity $\vec{v}$ and an electric charge of q_e experiences a force due to an electric field $\vec{E}$ and a magnetic field $\vec{B}$ as follows [3].

$$\vec{F} = q_e\big[\vec{E} + \vec{v} \times \vec{B}\big]$$
$$F_x = q_e E_x + v_y B_z - v_z B_y$$
$$F_y = q_e E_y + v_z B_x - v_x B_z$$
$$F_z = q_e E_z + v_x B_y - v_y B_x \qquad (2.4.1)$$

Here we consider the Lagrangian by taking the force associated with the magnetic field (the Lorenz force), which is complicated because the Lorenz force is proportional to the velocity of the charged object. The electric and magnetic fields are discussed using the voltage Φ and the vector potential $\vec{A}$ as follows:

$$\vec{E} = -\nabla\Phi - \frac{\partial \vec{A}}{\partial t} \left(E_q = -\frac{\partial \Phi}{\partial q} - \frac{\partial A_q}{\partial t} \quad q = x, y, z \right)$$
$$\vec{B} = \nabla \times \vec{A} \left(B_x = \frac{\partial A_z}{\partial y} - \frac{\partial A_y}{\partial z}, B_y = \frac{\partial A_x}{\partial z} - \frac{\partial A_z}{\partial x}, B_z = \frac{\partial A_y}{\partial x} - \frac{\partial A_x}{\partial y} \right) \qquad (2.4.2)$$

which is derived based on Maxwell's equations shown in chapter 3.

Taking $\vec{A} = \frac{1}{2}(-yB, xB, 0)$, $\vec{B} = (0,0,B)$, eq. (2.4.1) is rewritten as

$$F_x = q_e\big(E_x + v_y B\big)$$
$$F_y = q_e\big(E_y - v_x B\big)$$
$$F_z = q_e E_z \qquad (2.4.3)$$

Here we consider the formula of the Lagrangian as follows:

$$L_g = K - q_e\big(\Phi - \vec{v}\cdot\vec{A}\big) = \frac{m}{2}\big[v_x^2 + v_y^2 + v_z^2\big] - q_e\Phi + \frac{1}{2}q_e v_y x B - \frac{1}{2}q_e v_x y B$$
$$= \frac{m}{2}\big[v_x^2 + v_y^2 + v_z^2\big] - q_e\Phi + \frac{1}{2}q_e v_y A_y - \frac{1}{2}q_e v_x A_x \qquad (2.4.4)$$

Using this expression for the Lagrangian, the equation of motion is represented as:

$$\frac{d}{dt}\left(\frac{\partial L_g}{\partial v_x}\right) - \frac{\partial L_g}{\partial x} = m\frac{dv_x}{dt} - q_e E_x - q_e v_y B = m\frac{dv_x}{dt} - q_e\frac{\partial}{\partial x}\big(\Phi - v_y A_y\big) = 0$$
$$\frac{d}{dt}\left(\frac{\partial L_g}{\partial v_y}\right) - \frac{\partial L_g}{\partial y} = m\frac{dv_y}{dt} - q_e E_y + q_e v_x B = m\frac{dv_y}{dt} - q_e\frac{\partial}{\partial y}\big(\Phi - v_x A_x\big) = 0$$

$$\frac{d}{dt}\left(\frac{\partial L_g}{\partial v_z}\right) - \frac{\partial L_g}{\partial z} = m\frac{dv_z}{dt} - q_e E_z = m\frac{dv_x}{dt} - q_e\frac{\partial \Phi}{\partial z} = 0 \qquad (2.4.5)$$

which corresponds to eq. (2.4.3).

Using the momentum vector $\vec{p}$, eq.(2.4.4) can be rewritten as

$$L_g = \frac{1}{2m}|\vec{p}|^2 + \frac{1}{m}q_e\vec{p}\cdot\vec{A} - q_e\Phi$$
$$= \frac{1}{2m}\left[p_x^2 + p_y^2 + p_z^2\right] - q_e\Phi + \frac{1}{2m}q_e p_y xB - \frac{1}{2}q_e p_x yB$$

using the angular momentum $\vec{L} = \vec{r}\times\vec{p}$

$$= \frac{1}{2m}\left[p_x^2 + p_y^2 + p_z^2\right] - q_e\Phi + \frac{1}{2m}q_e L_z B = \frac{|\vec{p}|^2}{2m} - q\Phi + \frac{q_e}{2m}\vec{L}\cdot\vec{B} \qquad (2.4.6)$$

The magnetic dipole moment is defined as:

$$\overrightarrow{\mu_m} = \frac{q_e}{2m}\vec{L} \qquad (2.4.7)$$

and the energy given by the magnetic field is expressed as $\overrightarrow{\mu_m}\cdot\vec{B}$, which is called "Zeeman energy shift".

EXERCISE

The orbit of a mass m in the (x,y) directions (x: horizontal, y: vertical) is given by $y = y_0 + y_1\sin(ax)$.

There is a gravitational force $-mg$ in the y-direction. Derive the equation for the temporal change of x using the Lagrange equation.

(Answer)
Using eq. (2.3.5),
$$L_g = \frac{m}{2}[1 + [ay_1\cos(ax)]^2]\dot{x}^2 - mg[y_0 + y_1\sin(ax)]$$
$$\frac{d}{dt}\left(\frac{\partial L_g}{\partial \dot{x}}\right) - \frac{\partial L_g}{\partial x} = m[1 + [ay_1\cos(ax)]^2]\frac{d^2x}{dt^2} + mgay_1\cos(ax) = 0$$
$$\frac{d^2x}{dt^2} = -\frac{gay_1\cos(ax)}{1+[ay_1\cos(ax)]^2}$$

REFERENCES

[1] R. Chabey, and B. Sherwood, "Matter & interactions", *Modern Physics,* vol. 1, pp. 34-35, 2015.

[2] E. Davis, *Deriving kepler's law of planetary motion.* DavisEmily.pdf (uu.edu)

[3] P.G. Huray, *Maxwell's equations.* Wiley-IEEE, 2010, p. 22.
[http://dx.doi.org/10.1002/9780470549919]

CHAPTER 3

Fundamental Meaning and Typical Solutions of Maxwell's Equations

Abstract: The main objective is to understand the meaning of Maxwell's equations. It is a set of four differential equations that describe several fundamental laws that are already known. Why did these equations cause a revolution in physics? One reason is that the distribution of the electromagnetic field can be obtained by solving first- or second-order differential equations. Several examples have been shown to obtain the distribution of the electric field produced by simple electrodes. The three-dimensional trapping of charged matter can be achieved via two methods: the combination of a DC electric field and a DC magnetic field or using an AC electric field.

The most important aspect of Maxwell's equation is that it elucidates the identity of light as an electromagnetic wave. Light energy is given as the potential energy of electric and magnetic fields. Light has momentum and generates radiation pressure on a reflecting mirror.

The speed of light was determined to be independent of the observer, which was the basis for the theory of relativity. A fundamental aspect of the theory of special relativity is also introduced.

Keywords: Ampere's law, Coulomb's law, Cyclotron radiation, Electromagnetic wave, Faraday's law of induction, Gauss's theorem, Maxwell's equation, Lorenz transform, Stokes's theorem, The speed of light, Theory of relativity, Trap of charged matter.

3.1. WHAT ARE MAXWELL'S EQUATIONS?

The fundamentals of electromagnetism regarding an electric field $\vec{E}$ and magnetic field $\vec{B}$ are summarized by Maxwell's equations as follows (in SI units) [1]:

$\nabla \cdot \vec{E} = \frac{\rho}{\varepsilon}$ (*r*: the electric charge density, *e*: the permeability)

$$\frac{\partial E_x}{\partial x} + \frac{\partial E_y}{\partial y} + \frac{\partial E_z}{\partial z} = \frac{\rho}{\varepsilon} \qquad (3.1.1)$$

$$\nabla \times \vec{E} = -\frac{\partial \vec{B}}{\partial t}$$

$$\frac{\partial E_z}{\partial y} - \frac{\partial E_y}{\partial z} = -\frac{\partial B_x}{\partial t}, \frac{\partial E_x}{\partial z} - \frac{\partial E_z}{\partial x} = -\frac{\partial B_y}{\partial t}, \frac{\partial E_y}{\partial x} - \frac{\partial E_x}{\partial y} = -\frac{\partial B_z}{\partial t} \qquad (3.1.2)$$

$$\nabla \cdot \vec{B} = 0$$

$$\frac{\partial B_x}{\partial x} + \frac{\partial B_y}{\partial y} + \frac{\partial B_z}{\partial z} = 0 \qquad (3.1.3)$$

$\nabla \times \vec{B} = \mu \left[\vec{j} + \varepsilon \frac{\partial \vec{B}}{\partial t} \right]$ ($\vec{j}$: the electric current density, μ: the permittivity)

$$\frac{\partial B_z}{\partial y} - \frac{\partial B_y}{\partial z} = \mu \left[j_x + \varepsilon \frac{\partial E_x}{\partial t} \right], \frac{\partial B_x}{\partial z} - \frac{\partial B_z}{\partial x} = \mu \left[j_y + \varepsilon \frac{\partial E_y}{\partial t} \right], \frac{\partial B_y}{\partial x} - \frac{\partial B_x}{\partial y} = \mu \left[j_z + \varepsilon \frac{\partial E_z}{\partial t} \right] (3.1.4)$$

As shown in the following, these equations are novel expressions of laws that were already known.

Equation (3.1.1) is derived from Coulomb's law [2] for the electric field given by a point electric charge q_e

$$\vec{E} = \frac{q_e}{4\pi\varepsilon |\vec{r}|^3} \vec{r} \qquad (3.1.5)$$

The integral of $\vec{E}$ over the surface surrounding the electric charge is given as follows using polar coordinate ($x = r \sin\theta \cos\phi$, $y = r \sin\theta \sin\phi$, $z = r \cos\theta$):

$$\iint \vec{E} \cdot d\vec{S} = \iint |\vec{E}| r^2 \sin\theta \, d\theta \, d\phi = \frac{q_e}{\varepsilon} \qquad (3.1.6)$$

From Gauss's law [3], we have:

$$|\vec{E}| = \int \frac{\partial}{\partial r} |\vec{E}| dr$$

$$\iint \vec{E} \cdot d\vec{S} = \iiint \left[\frac{\partial |\vec{E}|}{\partial r} \right] dV = \frac{q_e}{\varepsilon} \qquad dV = r^2 \sin\theta \, dr \, d\theta \, d\phi$$

$$\text{using } \frac{\partial}{\partial r} = \frac{\partial x}{\partial r} \frac{\partial}{\partial x} + \frac{\partial y}{\partial r} \frac{\partial}{\partial y} + \frac{\partial z}{\partial r} \frac{\partial}{\partial z} = \sin\theta \cos\phi \frac{\partial}{\partial x} + \sin\theta \sin\phi \frac{\partial}{\partial y} + \cos\theta \frac{\partial}{\partial z}$$

$$E_x = \sin\theta \cos\phi |\vec{E}|, E_y = \sin\theta \sin\phi |\vec{E}|, E_z = \cos\theta |\vec{E}|$$

$$\frac{\partial |\vec{E}|}{\partial r} = \frac{\partial E_x}{\partial x} + \frac{\partial E_y}{\partial y} + \frac{\partial E_z}{\partial z} = \nabla \cdot \vec{E}$$

$$\iint \vec{E} \cdot d\vec{S} = \iiint \nabla \cdot \vec{E} \, dV = \frac{q_e}{\varepsilon} = \iiint \frac{\rho}{\varepsilon} \, dV \qquad (3.1.7)$$

and eq. (3.1.1) is derived. Since there is no magnetic charge, eq. (3.1.3) is derived from eq. (3.1.1) by changing $\vec{E} \rightarrow \vec{B}, \rho \rightarrow 0$.

Before explaining eqs. (3.1.2) and (3.1.4), Stokes's theorem [4] for the line integral of a vector quantity should be introduced. The line integral is a vector quantity, for which the x, y, and z-components result from the analyses of the yz, zx, and xy-planes, respectively. Here, we consider the component in the z-direction (line integral in the xy-plane). The line integral of the closed loop of a vector quantity $\vec{f}$ is given by

$$\oint \vec{f} \cdot d\vec{l}_z = \oint f_x dx + \oint f_y dy \tag{3.1.8}$$

Stokes's theorem shows that:

$$\oint \vec{f} \cdot d\vec{l}_z = \iint_{inside\ loop} \left[\frac{\partial f_y}{\partial x} - \frac{\partial f_x}{\partial y} \right] dS_z \quad dS_z = dxdy \tag{3.1.9}$$

The detailed derivation of eq. (3.1.9) is represented in Ref. [4]. As a simple case, we consider the line integral of a loop $(x_0, y_0) \rightarrow (x_1, y_0) \rightarrow (x_1, y_1) \rightarrow (x_0, y_1) \rightarrow (x_0, y_0)$. Then, we have:

$$\oint \vec{f} \cdot d\vec{l}_z = \int_{x_0}^{x_1} f_x(x, y_0)dx + \int_{y_0}^{y_1} f_y(x_1, y)dy + \int_{x_1}^{x_0} f_x(x, y_1)dx + \int_{y_1}^{y_0} f_y(x_0, y)dy$$

$$= \int_{x_0}^{x_1} [f_x(x, y_0) - f_x(x, y_1)]dx + \int_{y_0}^{y_1} [f_y(x_1, y) - f_y(x_0, y)]_y dy$$

$$= \int_{x_0}^{x_1} \int_{y_0}^{y_1} \left[\frac{\partial f_y(x,y)}{\partial x} - \frac{\partial f_x(x,y)}{\partial y} \right] dxdy \tag{3.1.10}$$

Next, we consider the vector quantity in the tangential direction of the circular loop, $\vec{f} = (f_x, f_y) = f(r)(-\sin\theta, \cos\theta)$, using the two-dimensional polar coordinate $(x = r\cos\theta, y = r\sin\theta)$. Then using:

$$\frac{\partial}{\partial x} = \frac{\partial r}{\partial x}\frac{\partial}{\partial r} + \frac{\partial \tan\theta}{\partial x}\frac{\partial\theta}{\partial \tan\theta}\frac{\partial}{\partial\theta} = \cos\theta\frac{\partial}{\partial r} - \frac{\sin\theta}{r}\frac{\partial}{\partial\theta}, \quad \frac{\partial}{\partial y} = \sin\theta\frac{\partial}{\partial r} + \frac{\cos\theta}{r}\frac{\partial}{\partial\theta} \tag{3.1.11}$$

$$\iint \left[\frac{\partial f_y}{\partial x} - \frac{\partial f_x}{\partial y} \right] d\vec{S}_z = \int_0^{2\pi} d\theta \int_0^r \left[\frac{\partial f(r)}{\partial r} + \frac{f(r)}{r} \right] rdr = 2\pi r f(r) = \oint \vec{f} \cdot d\vec{l}_z \tag{3.1.12}$$

is derived (see Fig. **3.1**).

Faraday's law of induction [5] shows that the voltage in the electric loop is given by

$$\Phi = \oint \vec{E}\,d\vec{l} = \iint_{inside\ loop} \frac{\partial \vec{B}}{\partial t} \cdot d\vec{S} \tag{3.1.13}$$

Using Stokes's law,

$$\oint \vec{E}\,d\vec{l} = \iint_{inside\ loop} [\nabla \times \vec{E}] \cdot d\vec{S} = \iint_{inside\ loop} \frac{\partial \vec{B}}{\partial t} \cdot d\vec{S} \tag{3.1.14}$$

and eq. (3.1.2) is derived.

Ampere's circuital law [6] states that:

$$\oint \vec{B}\,d\vec{l} = \iint_{inside\ loop} \mu \vec{j} \cdot d\vec{S} \tag{3.1.15}$$

However, there is a discrepancy in the case of a circuit containing a condenser for this law. From eq. (3.1.15), there should not be a magnetic field around the condenser. This discrepancy was solved using the constancy of the total electric charge:

$$\nabla \cdot \vec{j} + \frac{\partial \rho}{\partial t} = \nabla \cdot \left[\vec{j} + \varepsilon \frac{\partial \vec{E}}{\partial t} \right] = 0 \tag{3.1.16}$$

Using eq. (3.1.16), eq. (3.1.15) was corrected to (see Fig. **3.1**).

$$\oint \vec{B}\,d\vec{l} = \iint_{inside\ loop} \mu \left[\vec{j} + \varepsilon \frac{\partial \vec{E}}{\partial t} \right] \cdot d\vec{S} \tag{3.1.17}$$

and eq. (3.1.4) is derived using Stokes's law.

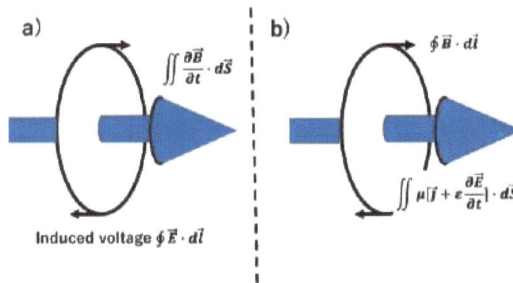

Fig. (3.1). a) Shows Faraday's law of induction. The line integral of the electric field is equal to the plane integral of the temporal change of the magnetic field. **b)** Shows the corrected Ampere's law.

The line integral of the magnetic field along a loop is given by the electric current and the temporal change of the electric field inside the loop.

3.2. WHY WERE MAXWELL'S EQUATIONS REVOLUTIONARY?

Maxwell's equations are simply a change in the expression of several laws related to electromagnetism. These equations made it possible to calculate the distribution of the electric or magnetic field by solving first- or second-order differential equations, making it possible to analyze the equations of motion of charged particles.

3.2.1. Can we Trap Charged Particles?

The equation of motion of a particle with an electric charge of q_e is given by,

$$m\frac{d^2\vec{r}}{dt^2} = q_e \left[\vec{E} + \frac{d\vec{r}}{dt} \times \vec{B} \right] \quad \text{shown in eq. (2.4.1)} \qquad (3.2.1)$$

First, we consider the possibility of trapping a charged particle at one point $(x, y, z) = (0,0,0)$ using an inhomogeneous DC electric field. In the space of $\rho = 0$, eq. (3.1.1) can be rewritten as:

$$\frac{\partial E_x}{\partial x} + \frac{\partial E_y}{\partial y} + \frac{\partial E_z}{\partial z} = 0 \qquad (3.2.2)$$

To trap in the x-direction, $q_e E_x > 0$ at $x < 0$ and $q_e E_x < 0$ at $x > 0$ is required, therefore, $q_e \frac{dE_x}{dx} < 0$ is required. To trap in all directions, $q_e \frac{dE_y}{dy} < 0$ and $q_e \frac{dE_z}{dz} < 0$ are also required. But eq. (3.2.2) shows that the signs of $\frac{dE_x}{dx}, \frac{dE_y}{dy}$ and $\frac{dE_z}{dz}$ cannot be the same. Therefore, charged particles cannot be trapped in all directions by a DC electric field.

The charged particle can be trapped using a DC electric field and a DC magnetic field, which is called the Penning trap [7]. When a magnetic field B_z is applied in the z-direction, the charged particles have a circular motion (cyclotron motion), as expected from the balance of the force given by the magnetic field (called the Lorentz force) and the centrifugal force:

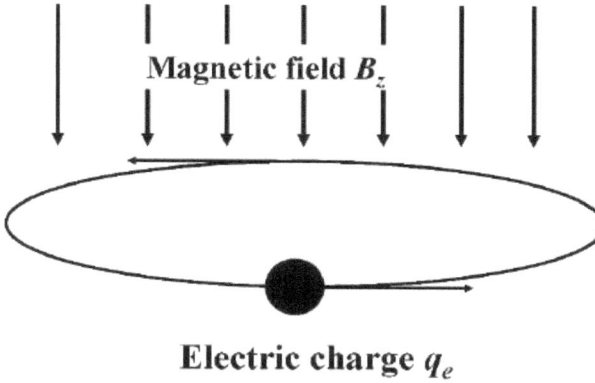

Fig. (3.2). Circular motion of a charged particle in the *xy*-plane under the influence of a magnetic field in the *z*-direction, called cyclotron motion.

$$mr\omega^2 = q_e r\omega B$$

$$\omega = \frac{q_e B_z}{m} \qquad \omega: \text{angular velocity} \qquad (3.2.3)$$

As a result of the cyclotron motion, charged particles can be trapped in a small region in the *xy*-plane, as shown in Fig. **(3.2)**. A DC electric field in the z-direction, that satisfies $q_e E_z > 0$ at $z < 0$ and $q_e E_z < 0$ at $z > 0$ is also applied so that the charged particle is trapped in the z-direction. The mass of the charged particle is determined by measuring the frequency of the cyclotron motion $\nu = \frac{\omega}{2\pi}$, and the mass of the charged particle can be determined. The masses of protons and antiprotons (chapter 5.5) have been confirmed to be equal with an uncertainty of 10^{-11} using this method [8].

Charged particles can be trapped using an inhomogeneous AC electric field. For an inhomogeneous AC electric field $\vec{E}(\vec{r}) \sin(\Omega t)$, trapping and expanding forces are periodically applied. The time-averaged force can be a trapping force in all directions under a certain condition. The equation of motion of a charged particle is given by:

$$m\frac{d^2\vec{r}}{dt^2} = q_e \vec{E}(\vec{r}) \sin(\Omega t) \qquad (3.2.4)$$

When the change in position within the period of the AC electric field $\delta \vec{r}$ is negligibly small compared to $|\vec{r}|$, the time-averaged force is obtained to be non-zero as follows (see Fig. **3.3**):

Micromotion is synchronized to the AC electric field

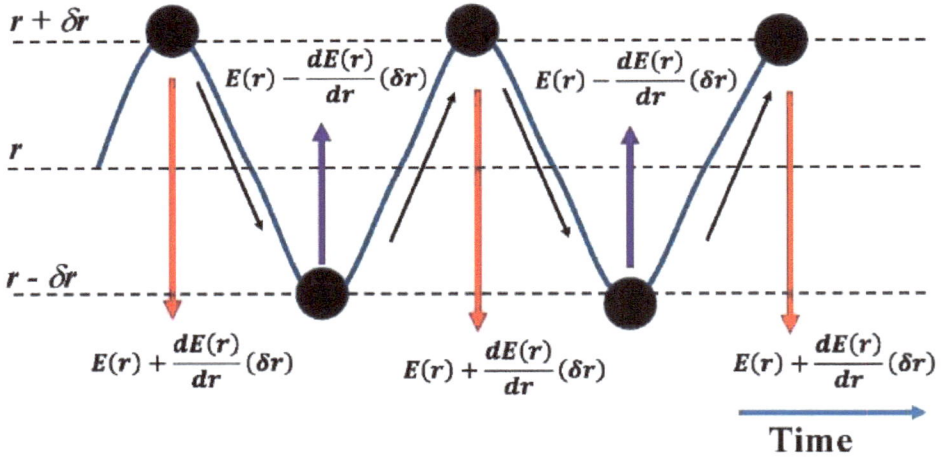

Fig. (3.3). The mechanism to trap a charged particle using an AC electric field. The micromotion of the charged particle is synchronized to the AC electric field and a non-zero averaged force is induced in one direction.

$$\vec{r} = \vec{r_0} + \delta \vec{r}$$

$$m \frac{d^2 \delta \vec{r}}{dt^2} = q_e \vec{E}(\vec{r_0}) \sin(\Omega t)$$

$$\delta \vec{r} = -\frac{q_e \vec{E}(\vec{r})}{m\Omega^2} \sin(\Omega t) + c_1 t + c_0$$

($c_{0,1}$: constant given by the initial condition)

$$\left[m \frac{d^2 \vec{r_0}}{dt^2} \right]_{ave} = \left[q_e \vec{E}(\vec{r_0} + \delta \vec{r}) \sin(\Omega t) \right]_{ave}$$

$$= \left[q_e \vec{E}(\vec{r_0}) \sin(\Omega t) \right]_{ave} + \left[q_e \frac{d\vec{E}(\vec{r_0})}{d\vec{r}} \delta \vec{r} \sin(\Omega t) \right]_{ave}$$

$$= -\left[q_e \frac{d\vec{E}(\vec{r_0})}{d\vec{r}} \left\{ \frac{q_e \vec{E}(\vec{r_0})}{m\Omega^2} \sin(\Omega t)^2 + (c_1 t + c_0) \sin(\Omega t) \right\} \right]_{ave} = -\frac{q_e^2 \vec{E}(\vec{r_0})}{2m\Omega^2} \frac{d\vec{E}(\vec{r_0})}{d\vec{r}} \quad \textbf{(3.2.5)}$$

The motion of the particle (change of r_0) as represented by eq. (3.2.5) is given as the motion with a pseudopotential field given by:

$$P_{ps}(\vec{r_0}) = \frac{|q_e \vec{E}(\vec{r_0})|^2}{4m\Omega^2}$$ (3.2.6)

Equation (3.2.6) shows $P_{ps}(\vec{r_0}) \geq 0$, therefore, the charged particle is trapped at the position where $|\vec{E_0}(\vec{r})|$ is minimum (for example, in the case of the hyperboloid electrode as shown in chapter 3.2.2). Note that

$$\left|\vec{E}(\vec{r_0})\right| \gg \left|\frac{d\vec{E}(\vec{r_0})}{d\vec{r}}\delta\vec{r}\right| \rightarrow \frac{q_c}{m\Omega^2}\left|\frac{d\vec{E}(\vec{r_0})}{d\vec{r}}\right| \ll 1$$ (3.2.7)

is required for the use of eq. (3.2.5). Ions can be trapped within a region smaller than 0.1 μm using this method, which is useful for the precision measurement of the transition frequencies of ions [9, 10].

3.2.2. Distribution of the Electric Field

The distribution of the electric field is obtained from eq. (3.2.2). However, it is often more convenient to consider the voltage Φ $(= -\int \vec{E} \cdot d\vec{r}$ when $\frac{\partial B}{\partial t} = 0)$ as follows:

$$\frac{\partial^2 \Phi}{\partial x^2} + \frac{\partial^2 \Phi}{\partial y^2} + \frac{\partial^2 \Phi}{\partial z^2} = 0$$ (3.2.8)

because the boundary condition is given by the voltage on the electrode surface. For the case of the hyperboloid electrode,

$$\begin{aligned} \Phi = \Phi_0 \quad & x^2 + y^2 - 2z^2 = -2z_0^2 \\ \Phi = 0 \quad & x^2 + y^2 - 2z^2 = r_0^2 \end{aligned}$$ (3.2.9)

the voltage distribution is given by:

$$\Phi(x, y, z) = -\frac{\Phi_0[x^2 + y^2 - 2z^2 - r_0^2]}{r_0^2 + 2z_0^2}$$ (3.2.10)

The uniqueness theorem [11] shows that eq. (3.2.10) is the only solution that satisfies eqs. (3.2.8) and (3.2.9). The electric field is given by:

$$E_x = \frac{2\Phi_0 x}{r_0^2 + 2z_0^2}, E_y = \frac{2\Phi_0 y}{r_0^2 + 2z_0^2}, \ E_z = -\frac{4\Phi_0 z}{r_0^2 + 2z_0^2} \tag{3.2.11}$$

Next, the electric field given by a cylindrical electrode is obtained as follows using the two-dimensional polar coordinates,

$$\frac{\partial^2 \Phi}{\partial x^2} + \frac{\partial^2 \Phi}{\partial y^2} = \frac{\partial^2 \Phi}{\partial r^2} + \frac{1}{r}\frac{\partial \Phi}{\partial r} + \frac{1}{r^2}\frac{\partial^2 \Phi}{\partial \theta} = 0$$

$$\Phi = \Phi_0 \quad r = r_0 \quad \text{(independent of } \theta\text{)} \tag{3.2.12}$$

Equation (3.2.12) is solved as:

$$\frac{\partial^2 \Phi}{\partial r^2} = -\frac{1}{r}\frac{\partial \Phi}{\partial r}$$

$$\int \frac{1}{\left(\frac{d\Phi}{dr}\right)} \frac{d\left(\frac{d\Phi}{dr}\right)}{dr} dr = -\int \frac{1}{r} dr \quad \ln\left(\frac{d\Phi}{dr}\right) = -\ln(r) + C$$

$$\Phi = \Phi_0 \left[1 - \ln\left(\frac{r}{r_0}\right)\right], \quad E_x = \frac{\Phi_0}{r}\cos\theta, \quad E_y = \frac{\Phi_0}{r}\sin\theta \tag{3.2.13}$$

It is generally difficult to obtain the solution to Maxwell's equation as an explicit formula. Different types of electromagnetic simulation software have been developed to perform numerical calculations of electric field's arbitrary arrangements of electrodes and circuits using the finite element method (calculation for the small region using simplified differential equations [12, 13].

3.3. ENERGY OF ELECTROMAGNETIC FIELDS

The energy *per* unit volume of an electric field in a vacuum is given by considering the work required to compress a charged sphere into a smaller radius $R \rightarrow R - \Delta R$ (volume $V \rightarrow V - \Delta V$). Using eq. (3.1.1),

$$W_{el} = 4\pi R^2 \int_R^{R-\Delta R} |\vec{E}| \rho dr = 4\pi R^2 \int_R^{R-\Delta R} |\vec{E}| \left[\varepsilon \frac{d|\vec{E}|}{dr}\right] dr = \frac{\varepsilon|\vec{E}|^2}{2} \Delta V$$

$$\Delta V = 4\pi R^2 \Delta R \tag{3.3.1}$$

The energy density of an electric field in a vacuum (volume of ΔV) associated with this work is given by $\frac{\varepsilon}{2}|\vec{E}|^2$.

The energy *per* unit volume of a magnetic field is obtained by considering the induction of the voltage Φ owing to the temporal change of the magnetic field in the z-direction B_z given by a circular current loop on the xy-plane with a radius of r_l. (see eq. (3.1.11))

$$\frac{\partial B_z}{\partial t}(\pi r_l^2) = \Phi \tag{3.3.2}$$

Considering the electric current density $\vec{J} = j_{0e}\frac{\vec{l}}{|\vec{l}|}$, the size of the cross-section of the conducting wire Δr (in the radial direction) and Δz (in the z direction), the energy generated by the current *per* unit time is given by:

$$W_{mg} = j_{0e}\Delta r\Delta z\Phi \tag{3.3.3}$$

For a circular loop with a radius of r_l, the magnetic field at the center is given by (eq. (3.1.15)),

$$B_z = \mu\frac{j_{0e}\Delta r}{2\pi r_l} \times 2\pi r_l = j_{0e}\Delta r\mu \tag{3.3.4}$$

and eq. (3.3.3) can be rewritten as:

$$W_{mg} = \frac{B_z}{\mu}\Phi\Delta z = \frac{B_z}{\mu}\frac{\partial B_d}{\partial t}(\pi r_l^2)\Delta z$$
$$= \frac{\partial}{\partial t}\left(\frac{B_z^2}{2\mu}\right)V \quad V = (\pi r_l^2)\Delta z \tag{3.3.5}$$

The energy density induced by the magnetic field is given by $\frac{1}{2\mu}\left|\vec{B}\right|^2$.

As shown in the next chapter, light is an electromagnetic wave, and its energy is stored in the electric and magnetic field components, as shown in the proceeding chapter.

3.4. ELECTROMAGNETIC WAVE AS THE IDENTITY OF LIGHT

Taking the AC electric field direction in the x-direction, eq. (3.1.2) with $\rho = 0$ and $\vec{J} = 0$ gives:

$$\frac{\partial E_x}{\partial z} = -\frac{\partial B_y}{\partial t}, \quad \frac{\partial E_x}{\partial y} = \frac{\partial B_z}{\partial t} \tag{3.4.1}$$

and eq. (3.1.4) can be expressed as:

$$\frac{\partial B_z}{\partial y} - \frac{\partial B_y}{\partial z} = -\varepsilon\mu\frac{\partial E_x}{\partial t} \tag{3.4.2}$$

The AC electric field in the *x*-direction induces an AC magnetic field in the *y*- or *z*-direction; here, we take the AC magnetic field in the *y*-direction. Thus, we have:

$$\frac{\partial E_x}{\partial z} = -\frac{\partial B_y}{\partial t} \rightarrow \frac{\partial^2 E_x}{\partial z^2} = -\frac{\partial^2 B_y}{\partial z\partial t}$$

$$\frac{\partial B_y}{\partial z} = -\varepsilon\mu\frac{\partial E_z}{\partial t} \rightarrow \frac{\partial^2 B_y}{\partial z\partial t} = -\varepsilon\mu\frac{\partial^2 E_x}{\partial t^2}$$

$$\frac{\partial^2 E_x}{\partial z^2} = \varepsilon\mu\frac{\partial^2 E_x}{\partial t^2} \tag{3.4.3}$$

To solve eq. (3.4.3), E_x is assumed to be the product of the functions Z_E and T_E, which depend only on *z* and *t*, respectively. Then we have:

$$T_E\frac{\partial^2 Z_E}{\partial z^2} = \varepsilon\mu Z_E\frac{\partial^2 T_E}{\partial t^2}$$

$$\frac{1}{\varepsilon\mu}\frac{1}{Z_E}\frac{\partial^2 Z_E}{\partial z^2} = \frac{1}{T_E}\frac{\partial^2 T_E}{\partial t^2} = C$$

$$T_E = a_+e^{\sqrt{C}t} + a_-e^{-\sqrt{C}t} \quad Z_E = b_+e^{\sqrt{\varepsilon\mu C}z} + b_-e^{-\sqrt{\varepsilon\mu C}t} \tag{3.4.4}$$

When $C > 0$, E_x is an expanding or damping function of *t* and *z*. When $C < 0$ ($C = -\Omega^2$),

$$E_x = c_1e^{i\Omega(t+\sqrt{\varepsilon\mu}z)} + c_2e^{i\Omega(t-\sqrt{\varepsilon\mu}z)} + c_3e^{-i\Omega(t+\sqrt{\varepsilon\mu}z)} + c_4e^{-i\Omega(t-\sqrt{\varepsilon\mu}z)}$$

$$= E_0\sin[\Omega(t \pm \sqrt{\varepsilon\mu}z) + \varphi_0] \tag{3.4.5}$$

Equation (3.4.5) shows that E_x is a waveform that propagates in the *z*-direction with a velocity of $c_m = \mp\frac{1}{\sqrt{\varepsilon\mu}}$. For B_y, we have:

$$\frac{\partial E_x}{\partial z} = -\frac{\partial B_y}{\partial t}, \frac{\partial B_y}{\partial z} = -\varepsilon\mu\frac{\partial E_z}{\partial t} \rightarrow B_y = B_0\sin[\Omega(t \pm \sqrt{\varepsilon\mu}z) + \varphi_0] \quad B_0 = -\sqrt{\varepsilon\mu}E_0 \tag{3.4.6}$$

The estimated propagation speed $c_m = \frac{1}{\sqrt{\varepsilon\mu}}$ corresponds to the speed of light, as measured by Fizeau and Foucault [14]. Therefore, the light is proven to be an electromagnetic wave. A change in the electric field induces a change in the magnetic field and vice versa. The change in the electric and magnetic fields propagates as waves. The directions of the electric field, magnetic field, and propagation are orthogonal to each other. In a propagating wave, the phases of the oscillation of the electric and magnetic fields are equal.

The light intensity (energy *per* unit time) is given by:

$$I_{light} = c_m S\big[\langle P_{el}\rangle_{ave} + \langle P_{mg}\rangle_{ave}\big]$$

$$\langle P_{el}\rangle_{ave} = \frac{\varepsilon E_0^2}{4}, \quad \langle P_{mg}\rangle_{ave} = \frac{B_0^2}{4\mu} \quad S: \text{area of the light spot} \qquad \textbf{(3.4.7)}$$

From eq. (3.4.6), $\langle P_{el}\rangle_{ave} = \langle P_{mg}\rangle_{ave}$ is obtained.

When light is reflected by a mirror, the electric field on the surface of the mirror is zero, and the phase of the electric field in the reflected light changes by π ($E_0 \rightarrow -E_0$). The phase of the magnetic field does not change by the reflection ($B_0 \rightarrow B_0$). From eq. (3.1.15), an electric current of $j_x = \frac{B_y}{\mu}$ is induced on the surface of the mirror (the magnetic field is zero inside the mirror). Taking $z = 0$ as the mirror position, the standing wave formed by the interference between the incident and reflected waves is given by

$$
\begin{aligned}
E_x &= E_0 \sin[\Omega(t + \sqrt{\varepsilon\mu}z) + \varphi_0] - E_0 \sin[\Omega(t - \sqrt{\varepsilon\mu}z) + \varphi_0] \\
&= 2E_0 \cos[\Omega t + \varphi_0] \sin[\Omega\sqrt{\varepsilon\mu}z] \\
B_y &= B_0 \sin[\Omega(t + \sqrt{\varepsilon\mu}z) + \varphi_0] + B_0 \sin[\Omega(t - \sqrt{\varepsilon\mu}z) + \varphi_0] \\
&= 2B_0 \sin[\Omega t + \varphi_0] \cos[\Omega\sqrt{\varepsilon\mu}z] \qquad \textbf{(3.4.8)}
\end{aligned}
$$

Averaging by time, the energy density given by:

$$P_{light} = \langle\frac{\varepsilon E_x^2}{2}\rangle_{ave} + \langle\frac{B_y^2}{2\mu}\rangle_{ave} = \varepsilon E_0^2 \qquad \textbf{(3.4.9)}$$

When light is reflected by a mirror, the mirror experiences a radiation pressure; the Lorenz force between the electric current and the magnetic field of $2B_y$ (sum of the field of the incident and reflected waves) is given by:

$$F_{radiation} = S\langle j_x B_y\rangle_{ave} = S\langle \frac{B_y^2}{\mu}\rangle_{ave} = \frac{2}{c_m} I_{light} \tag{3.4.10}$$

We can consider that light has a momentum p_{light} in the propagation direction. The change in the momentum $p_{light} \to -p_{light}$ occurs during reflection. The change in momentum *per* unit time results in the radiation pressure.

3.5. DESCRIPTION OF MAXWELL'S EQUATIONS USING THE VECTOR POTENTIAL

For complicated analysis, it is useful to simplify Maxwell's equations using a vector potential $\vec{A}$ that satisfies:

$$\vec{B} = \nabla \times \vec{A} \tag{3.5.1}$$

The solution of $\vec{A}$ must exist since eq. (3.1.3) holds for $\vec{B}$. From eq. (3.1.2),

$$\vec{E} = -\frac{\partial \vec{A}}{\partial t} - \nabla\Phi \quad (\nabla \times \nabla\Phi = 0) \tag{3.5.2}$$

is derived. The voltage Φ is also called the scalar potential in comparison with the vector potential. Equation (3.1.1) can be rewritten as:

$$\nabla^2\Phi + \nabla \cdot \frac{\partial \vec{A}}{\partial t} = -\frac{\rho}{\varepsilon} \tag{3.5.3}$$

Equation (3.1.4) can be rewritten as:

$$\nabla \times \nabla \times \vec{A} = \mu\left[\vec{j} - \varepsilon\left(\frac{\partial^2 \vec{A}}{\partial t^2} + \frac{\partial(\nabla\Phi)}{\partial t}\right)\right]$$

$$\text{using } \nabla \times \nabla \times \vec{A} = \nabla(\nabla \cdot \vec{A}) - \nabla^2\vec{A}$$

$$\left[\nabla^2 - \varepsilon\mu\frac{\partial^2}{\partial t^2}\right]\vec{A} - \nabla\left(\nabla \cdot \vec{A} + \varepsilon\mu\frac{\partial\Phi}{\partial t}\right) = -\mu\vec{j} \tag{3.5.4}$$

The solutions of $\vec{A}$ and Φ are not unique, and the values of $\vec{E}$ and $\vec{B}$ do not change with the following transformation:

$$\vec{A} \to \vec{A} + \nabla\chi \quad \Phi \to \Phi - \frac{\partial\chi}{\partial t}$$

$$\nabla \times (\vec{A} + \nabla\chi) = \nabla \times \vec{A} = \vec{B}$$

$$-\nabla\left(\Phi - \frac{\partial\chi}{\partial t}\right) - \frac{\partial}{\partial t}(\vec{A} + \nabla\chi) = -\nabla\Phi - \frac{\partial}{\partial t}\vec{A} = \vec{E} \tag{3.5.5}$$

Using an appropriate gauge (called the Lorenz gauge) satisfying:

$$\nabla \cdot \vec{A} + \varepsilon\mu\frac{\partial\Phi}{\partial t} = 0 \quad \left(\nabla \cdot \frac{\partial\vec{A}}{\partial t} = -\varepsilon\mu\frac{\partial^2\Phi}{\partial t^2}\right) \tag{3.5.6}$$

eqs. (3.5.3) and (3.5.4) can be simplified to:

$$\nabla^2\Phi - \varepsilon\mu\frac{\partial^2\Phi}{\partial t^2} = -\frac{\rho}{\varepsilon} \tag{3.5.7}$$

$$\nabla^2\vec{A} - \varepsilon\mu\frac{\partial^2 A}{\partial t^2} = -\mu\vec{j} \tag{3.5.8}$$

When $\rho = 0$ $\vec{j} = 0$, eqs. (3.5.7) and (3.5.8) are the equations that describe wave propagation. Considering the propagation time,

$$\Phi(\vec{r}, t) = \frac{1}{4\pi\varepsilon} \iiint \frac{\rho\left(\vec{r}\prime, t - \frac{|\vec{r} - \vec{r}\prime|}{c_m}\right)}{|\vec{r} - \vec{r}\prime|} dV' \tag{3.5.9}$$

$$\vec{A}(\vec{r}, t) = \frac{\mu}{4\pi} \iiint \frac{\vec{j}\left(\vec{r}\prime, t - \frac{|\vec{r} - \vec{r}\prime|}{c_m}\right)}{|\vec{r} - \vec{r}\prime|} dV' \tag{3.5.10}$$

Here, we consider the distribution of the electromagnetic field when a particle with an electric charge of q_e has a rotational motion $(x, y, z) = (r_0 \cos(\omega t), r_0 \sin(\omega t), 0)$, which results in a current of $(-\omega r_0 \sin(\omega t), \omega r_0 \cos(\omega t), 0)$. Assuming $|\vec{r}| \gg r_0$, the components of $\vec{A}$ are given by

$$A_x(\vec{r}, t) = \frac{\mu}{4\pi} \frac{-q_e\omega r_0 \sin\left[\omega\left(t - \frac{|\vec{r}|}{c_m}\right)\right]}{|\vec{r}|} \quad A_y(\vec{r}, t) = \frac{\mu}{4\pi} \frac{q_e\omega r_0 \cos\left[\omega\left(t - \frac{|\vec{r}|}{c_m}\right)\right]}{|\vec{r}|}$$

$$A_z(\vec{r}, t) = 0 \qquad\qquad (3.5.11)$$

From eq. (3.5.6),

$$\frac{\partial \Phi(\vec{r},t)}{\partial t} = -\frac{1}{\varepsilon\mu}\left[\frac{\partial A_x}{\partial x} + \frac{\partial A_y}{\partial y} + \frac{\partial A_z}{\partial z}\right]$$

$$= \frac{q_e r_0}{4\pi\varepsilon}\left[\frac{\omega\{-x\sin\left[\omega\left(t-\frac{|\vec{r}|}{c_m}\right)\right]+y\cos\left[\omega\left(t-\frac{|\vec{r}|}{c_m}\right)\right]\}}{|r|^3} + \frac{\omega^2\{-x\cos\left[\omega\left(t-\frac{|\vec{r}|}{c_m}\right)\right]-y\sin\left[\omega\left(t-\frac{|\vec{r}|}{c_m}\right)\right]\}}{|r|^2 c_m}\right] \qquad (3.5.12)$$

is obtained and

$$\Phi(\vec{r},t) = \frac{q_e r_0}{4\pi\varepsilon}\left[\frac{\{x\cos\left[\omega\left(t-\frac{|\vec{r}|}{c_m}\right)\right]+y\sin\left[\omega\left(t-\frac{|\vec{r}|}{c_m}\right)\right]\}}{|r|^3} + \frac{\omega\{-x\sin\left[\omega\left(t-\frac{|\vec{r}|}{c_m}\right)\right]+y\cos\left[\omega\left(t-\frac{|\vec{r}|}{c_m}\right)\right]\}}{|r|^2 c_m} + \frac{1}{|\vec{r}| r_0}\right] \qquad (3.5.13)$$

is given. The third term of eq. (3.5.13) is the time independent term, which gives the DC voltage. The circular motion induces the radiation with a frequency of $\nu = \frac{\omega}{2\pi}$. The second term of eq. (3.5.13) is induced by the delay in determining the influence of the motion of the charged particle. At a long distance, the first term is negligibly small in comparison with the second term. The magnetic field is given by

$$B_x = \frac{\partial A_z}{\partial y} - \frac{\partial A_y}{\partial z} = \frac{\mu}{4\pi}\frac{q_e\omega r_0 z\cos\left[\omega\left(t-\frac{|\vec{r}|}{c_m}\right)\right]}{|\vec{r}|^3} + \frac{\mu}{4\pi}\frac{q_e\omega^2 r_0 z\sin\left[\omega\left(t-\frac{|\vec{r}|}{c_m}\right)\right]}{|\vec{r}|^2 c_m}$$

$$B_y = \frac{\partial A_x}{\partial z} - \frac{\partial A_z}{\partial x} = \frac{\mu}{4\pi}\frac{q_e\omega r_0 z\sin\left[\omega\left(t-\frac{|\vec{r}|}{c_m}\right)\right]}{|\vec{r}|^3} + \frac{\mu}{4\pi}\frac{q_e\omega^2 r_0 z\cos\left[\omega\left(t-\frac{|\vec{r}|}{c_m}\right)\right]}{|\vec{r}|^2 c_m}$$

$$B_z = \frac{\partial A_y}{\partial x} - \frac{\partial A_x}{\partial y}$$

$$= \frac{\mu}{4\pi}\frac{q_e\omega r_0\{x\cos\left[\omega\left(t-\frac{|\vec{r}|}{c_m}\right)\right]+y\sin\left[\omega\left(t-\frac{|\vec{r}|}{c_m}\right)\right]\}}{|\vec{r}|^3} + \frac{\mu}{4\pi}\frac{q_e\omega^2 r_0\{x\sin\left[\omega\left(t-\frac{|\vec{r}|}{c_m}\right)\right]-y\cos\left[\omega\left(t-\frac{|\vec{r}|}{c_m}\right)\right]\}}{|\vec{r}|^2 c_m} \qquad (3.5.14)$$

The second terms are induced by the delay in the influence of the motion of the charged particle and dominates at long distances. The electric field of radiation at a long distance is given by:

$$\vec{E} = -\nabla\Phi - \frac{\partial \vec{A}}{\partial t}$$

$$E_x = \frac{q_e r_0 \omega}{4\pi\varepsilon}\left[\frac{(y^2+z^2-x^2)\sin\left[\omega\left(t-\frac{|\vec{r}|}{c_m}\right)\right]+2xy\cos\left[\omega\left(t-\frac{|\vec{r}|}{c_m}\right)\right]}{|r|^4 c_m} + \frac{\omega\cos\left[\omega\left(t-\frac{|\vec{r}|}{c_m}\right)\right]}{|\vec{r}| c_m^2}\right]$$

$$E_y = \frac{q_e r_0 \omega}{4\pi\varepsilon}\left[\frac{-2xy\sin\left[\omega\left(t-\frac{|\vec{r}|}{c_m}\right)\right]+(x^2+z^2-y^2)\cos\left[\omega\left(t-\frac{|\vec{r}|}{c_m}\right)\right]}{|r|^4 c_m}+\frac{\omega\sin\left[\omega\left(t-\frac{|\vec{r}|}{c_m}\right)\right]}{|\vec{r}|c_m^2}\right]$$

$$E_z = \frac{q_e r_0 \omega}{4\pi\varepsilon}\left[\frac{-x\,z\sin\left[\omega\left(t-\frac{|\vec{r}|}{c_m}\right)\right]+yz\cos\left[\omega\left(t-\frac{|\vec{r}|}{c_m}\right)\right]}{|r|^4 c_m}\right] \qquad (3.5.15)$$

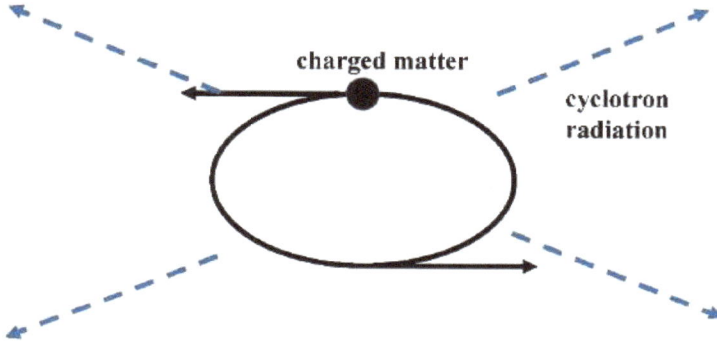

Fig. (3.4). Cyclotron radiation induced by the circular motion of a charged matter.

For a long distance, $|\vec{E}|$ and $|\vec{B}|$ are proportional to $\frac{r_0\omega^2}{|\vec{r}|}$, and are derived from the delay in the influence of the change in the electric magnetic field due to the motion of the charged body. The motion of the charged particle induces a change of the electric and magnetic fields, which propagate as electromagnetic waves. The energy expanded as radiation energy was supplied from the motion energy of the charged particle. The radiation in the direction parallel to the motion of the charged particle gives the applied deceleration force because the radiation has momentum in the propagation direction (see chapter 3.4.10). The radiation induced by the circular motion is called cyclotron radiation (see Fig. **3.4**).

3.6. SIMPLE INTERPRETATION OF THE THEORY OF SPECIAL RELATIVITY

Maxwell's equations facilitated the discovery of the identity of light. However, the speed of light propagating in a medium $c_m = \frac{1}{\sqrt{\varepsilon\mu}}$ is the speed against the medium. For propagation in a vacuum, what is the medium that serves as the standard for the speed of light? The values of ε and μ in vacuum space (ε_0 and μ_0) are universal constants and the speed of light in a vacuum $c = \frac{1}{\sqrt{\varepsilon_0\mu_0}}$ does not depend on the motion of the observer. However, the speed of light was expected to depend on the motion of the observer. The idea of ether as the propagation medium of light was proposed. The speed of light in a vacuum can then be defined as the speed against the ether. If we move against the ether, the speed of light is expected to shift. However, this effect was estimated to be too small to be detected. The speed of light was measured as the propagation time of a round trip to and from a mirror. When the observer moves against the ether with a velocity v,

$$T_p = \frac{L_b}{c+v} + \frac{L_b}{c-v} = \frac{2L_b c}{c^2 - v^2} \approx \frac{2L_b}{c}\left[1 + \left(\frac{v}{c}\right)^2\right]$$

L_b: distance to the mirror   **(3.6.1)**

It was not realistic to expect the influence of the revolution of the earth ($v = 10^{-4}c$) by direct measurement of the speed of light, given that a fractional measurement uncertainty lower than 10^{-8} is required. However, Michelson and Morley expected that this effect could be detected using an interferometer to combine light propagated in directions parallel and perpendicular to the Earth's orbital and rotational motion after reflection by mirrors [15]. Using an interferometer and light with a frequency of v, the change in the difference in the propagation time of δt was detected based on the change in the phase difference of $\delta\varphi = 2\pi v(\delta t)$. For a frequency of 10^{14} Hz, a change in the propagation time by 10^{-15} s was detected by the change in the phase difference of the interfering light beams (0.6 radians), and the effect of the Earth's orbital and rotational motion was expected to be detectable. However, the speed of light was measured to be constant within a fractional uncertainty of 10^{-8}. Using a laser source, the constancy of the speed of light has been confirmed with an uncertainty of 10^{-15} [16]. The speed of light in a vacuum is a universal value (in 1983 defined as 299792458 m/s) independent of the motion of the observers. The theory of relativity [17] was established based on the constancy of the speed of light in a vacuum.

We consider the coordinates (x,y,z) and (x',y',z'); (x',y',z') are coordinates of an observer moving with a velocity of v on the x-direction, as shown in Fig. (**3.5**). For classical mechanics, the relationship between both coordinates is given by:

$$x' = x - vt, y' = y, z' = z, t' = t$$
                                                                                   (3.6.2)

and the speed of light changes with $c' = c - v$. Another formula is required for the translation between different coordinates while maintaining a constant speed of light. The fundamentals of the special theory of relativity, for which $(ct')^2 = x'^2 + y'^2 + z'^2$ holds when $(ct)^2 = x^2 + y^2 + z^2$. Using a simple idea, $y' = y$ and $z' = z$ are assumed. Then $ct' = x'$ holds when $ct = x$. We consider the relationship between (x', ct') and (x, y) for the matrix:

$$\begin{pmatrix} x' \\ ct' \end{pmatrix} = \begin{pmatrix} p & q \\ r & s \end{pmatrix} \begin{pmatrix} x \\ ct \end{pmatrix} \tag{3.6.3}$$

with the requirement

$$ct' = x' \rightarrow rx + sct = px + qct , \quad (s-q)ct = (p-r)x \rightarrow s - q = p - r$$
$$q = -p\frac{v}{c}, s = p \quad \text{(with } v \ll c, \text{ eq. (3.6.1))} \rightarrow r = -\frac{v}{c}p$$

$$\begin{pmatrix} x' \\ ct' \end{pmatrix} = p \begin{pmatrix} 1 & -\frac{v}{c} \\ -\frac{v}{c} & 1 \end{pmatrix} \begin{pmatrix} x \\ ct \end{pmatrix} \tag{3.6.4}$$

For $v \rightarrow -v$, the matrix is transformed to an inverse matrix.

$$p^2 \begin{pmatrix} 1 & \frac{v}{c} \\ \frac{v}{c} & 1 \end{pmatrix} \begin{pmatrix} 1 & -\frac{v}{c} \\ -\frac{v}{c} & 1 \end{pmatrix} = \begin{pmatrix} 1 & 0 \\ 0 & 1 \end{pmatrix} \quad p^2 = \frac{1}{1-\left(\frac{v}{c}\right)^2}$$

$$\begin{pmatrix} x' \\ ct' \end{pmatrix} = \frac{1}{\sqrt{1-\left(\frac{v}{c}\right)^2}} \begin{pmatrix} 1 & -\frac{v}{c} \\ -\frac{v}{c} & 1 \end{pmatrix} \begin{pmatrix} x \\ ct \end{pmatrix} \tag{3.6.5}$$

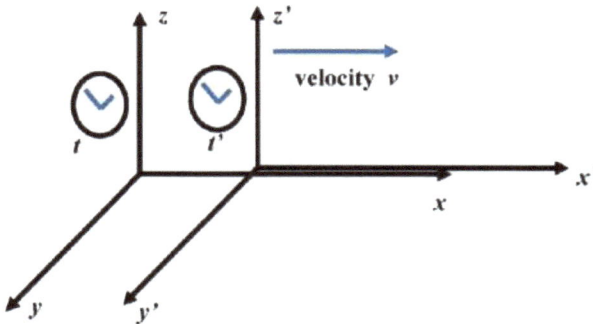

Fig. (3.5). Coordinates (x,y,z) and (x',y',z'), where (x',y',z') is the coordinate for an observer. moving with the velocity of v on the x-direction.

The transformation of the four-dimensional vector (x, y, z, ct) is given by:

$$\begin{pmatrix} x' \\ y' \\ z' \\ ct' \end{pmatrix} = \begin{pmatrix} \dfrac{1}{\sqrt{1-(v/c)^2}} & 0 & 0 & \dfrac{-(v/c)}{\sqrt{1-(v/c)^2}} \\ 0 & 1 & 0 & 0 \\ 0 & 0 & 1 & 0 \\ \dfrac{-(v/c)}{\sqrt{1-(v/c)^2}} & 0 & 0 & \dfrac{1}{\sqrt{1-(v/c)^2}} \end{pmatrix} \begin{pmatrix} x \\ y \\ z \\ ct \end{pmatrix} \tag{3.6.6}$$

which is called the "Lorenz transformation".

What happens to the length and the time interval in a moving coordinate system?

$$x'_1 - x'_2 = \frac{1}{\sqrt{1-\left(\frac{v}{c}\right)^2}}\left[(x_1 - vt_1) - (x_2 - vt_2)\right]$$

$$ct'_1 - ct'_2 = \frac{1}{\sqrt{1-\left(\frac{v}{c}\right)^2}}\left[\left(-\frac{v}{c}x_1 + ct_1\right) - \left(-\frac{v}{c}x_2 + ct_2\right)\right] \tag{3.6.7}$$

The length in the moving coordinate system should be considered for the following:

$$t'_1 = t'_2 \qquad t_1 - t_2 = \frac{v}{c^2}(x_1 - x_2)$$

and

$$x'_1 - x'_2 = \sqrt{1 - \left(\frac{v}{c}\right)^2}(x_1 - x_2) \tag{3.6.8}$$

is derived. The time interval for $x'_1 = x'_2$ is

$$t'_1 - t'_2 = \sqrt{1 - \left(\frac{v}{c}\right)^2}(t_1 - t_2) \tag{3.6.9}$$

We can observe particles with short lifetimes in cosmic rays from distant places (several light years) [18]. This is because particles move at high velocities (close to the speed of light), and time slows down in the moving coordinates system based on the motion of particles.

The velocity component (V_x, V_y, V_z) in the moving coordinate system is given by:

$$V'_x = \frac{dx\prime}{dt\prime} = \frac{d[\gamma(x-vt)]}{d\left[\gamma\left(t-\frac{vx}{c^2}\right)\right]} \qquad \left(\gamma = \frac{1}{\sqrt{1-\left(\frac{v}{c}\right)^2}}\right)$$

taking $x = V_x t$

$$= \frac{d[\gamma(x-vt)]}{d\left[\gamma\left(t-\frac{vV_x}{c^2}t\right)\right]} = \frac{V_x-v}{1-\frac{vV_x}{c^2}}$$

$$V'_y = \frac{dy\prime}{dt\prime} = \frac{dy}{d\left[\gamma\left(t-\frac{vV_x}{c^2}t\right)\right]} = \frac{1}{\gamma}\frac{V_y}{1-\frac{vV_x}{c^2}}$$

$$V'_z = \frac{1}{\gamma}\frac{V_z}{1-\frac{vV_x}{c^2}} \tag{3.6.10}$$

Here, we consider the case $V_x = c \cos \theta$ and $V_y = c \sin \theta$. Then

$$V'_x = \frac{c\cos\theta-v}{1-\frac{v}{c}\cos\theta}, \quad V'_y = \sqrt{1-\left(\frac{v}{c}\right)^2}\frac{c\sin\theta}{1-\frac{v}{c}\cos\theta}$$

$$V'^2_x + V'^2_y = c^2 \tag{3.6.11}$$

and the speed of light is constant for any moving coordinate system. The Lorenz transformation was originally considered by Lorenz, taking v as the speed against the ether to explain the constancy of the speed of light. According to the theory of relativity, v is the relative velocity. When A observes B moving with a relative velocity of v, B observes A moving with a relative velocity of -v. From the theory of relativity, A and B observe each other with equally shorter lengths and slower running time. However, this interpretation is not valid based on Lorentz's theory. When the velocity of A against the ether is v_0, the velocity of B against the ether is $v_0 + v$. For the theory of relativity, all physical laws in a coordinate system are also valid for Lorentz transformed coordinate systems (Lorentz invariance), as shown in the following. However, the Lorenz invariance is not valid in Lorentz's theory.

For classical mechanics, momentum is defined as $\vec{p} = m\vec{V}$. The conservation of momentum   and   energy E $(\vec{p_1} + \vec{p_2} = \vec{p_3} + \vec{p_4}$, $E_1 + E_2 = E_3 + E_4)$ is fundamental to mechanics, and we assume that it holds in a coordinate system. However, the conservation of momentum does not hold, considering $\vec{p'} = m\vec{V'}$ shown in Eq. (3.6.8). We need another definition of momentum that converges to the momentum of classical mechanics for $|\vec{V}| \ll c$. If the four-dimensional momentum vectors $\overrightarrow{p^4} = \left(p_x, p_y, p_z, {E}/{c}\right)$ are transformed into a moving coordinate system using the Lorenz transformation:

$$\begin{pmatrix} p'_x \\ p'_y \\ p'_z \\ E'/c \end{pmatrix} = \begin{pmatrix} \frac{1}{\sqrt{1-(v/c)^2}} & 0 & 0 & \frac{-(v/c)}{\sqrt{1-(v/c)^2}} \\ 0 & 1 & 0 & 0 \\ 0 & 0 & 1 & 0 \\ \frac{-(v/c)}{\sqrt{1-(v/c)^2}} & 0 & 0 & \frac{1}{\sqrt{1-(v/c)^2}} \end{pmatrix} \begin{pmatrix} p_x \\ p_y \\ p_z \\ E/c \end{pmatrix} \qquad (3.6.12)$$

the conservation of the four-dimensional momentum vector in this system guarantees conservation in another coordinate system. We consider the transformation from $(0,0,0, E/c)$,

$$p'_x = \frac{-\frac{v}{c^2}E}{\sqrt{1-\left(\frac{v}{c}\right)^2}} = \frac{\frac{V'_x}{c^2}E}{\sqrt{1-\left(\frac{V'_x}{c}\right)^2}} \qquad (3.6.13)$$

$$E' = \frac{E}{\sqrt{1-\left(\frac{V'_x}{c}\right)^2}} \qquad (3.6.14)$$

Taking the rest energy $E = mc^2$,

$$p'_x = \frac{mV'_x}{\sqrt{1-\left(\frac{V'_x}{c}\right)^2}} \qquad (3.6.15)$$

$$E' = \frac{mc^2}{\sqrt{1-\left(\frac{V'_x}{c}\right)^2}} \qquad (3.6.16)$$

For $V'_x \ll c$, eqs. (3.6.15) and (3.6.16) are approximated to:

$$p'_x = mV'_x \qquad (3.6.17)$$

and

$$E' = mc^2 + \frac{m}{2}V'^2_x \qquad (3.6.18)$$

respectively. Equation (3.6.17) shows the momentum for classical mechanics and eq. (3.6.18) denotes the sum of the rest energy and kinetic energy in the case of classical mechanics (eq. (2.1.6)). We consider the transformation from $(p_x, 0, 0, E/c)$. From eq. (3.6.12),

$$p'_x = \frac{p_x - vE/c^2}{\sqrt{1-(v/c)^2}}$$

$$E'/c^2 = \frac{E/c^2 - vp_x/c^2}{\sqrt{1-(v/c)^2}} \tag{3.6.19}$$

is derived. Equation (3.6.19) is confirmed using:

$$p_x = \frac{mV_x}{\sqrt{1-\frac{V_x^2}{c^2}}} \quad E = \frac{mc^2}{\sqrt{1-\frac{V_x^2}{c^2}}}$$

$$p'_x = \frac{mV'_x}{\sqrt{1-\frac{V'^2_x}{c^2}}} = \frac{m(V_x - v)}{\left(1-\frac{V_x v}{c^2}\right)\sqrt{1-\left[\frac{V_x-v}{c^2-V_x v}\right]^2}} = \frac{m(V_x - v)}{\sqrt{1-(V_x/c)^2}\sqrt{1-(v/c)^2}} = \frac{p_x - vE/c^2}{\sqrt{1-(v/c)^2}}$$

$$E' = \frac{mc^2}{\sqrt{1-\frac{V'^2_x}{c^2}}} = \frac{mc^2}{\sqrt{1-\left[\frac{V_x-v}{c^2-V_x v}\right]^2}} = \frac{E/c^2 - vp_x/c^2}{\sqrt{1-(v/c)^2}} \left(\text{using } V'_x = \frac{V_x-v}{1-\frac{V_x v}{c^2}}\right) \tag{3.6.20}$$

Equation (3.6.12) shows $p'_{y,z} = p_{y,z}$, which is confirmed from the general formula for momentum and energy as

$$p'_{x,y,z} = \frac{mV'_{x,y,z}}{\sqrt{1-\frac{V'^2_x + V'^2_y + V'^2_z}{c^2}}} \tag{3.6.21}$$

$$E' = \frac{mc^2}{\sqrt{1-\frac{V'^2_x + V'^2_y + V'^2_z}{c^2}}} \tag{3.6.22}$$

Comparing

$$p_y = \frac{mV_y}{\sqrt{1-\frac{V_y^2}{c^2}}} \qquad p'_y = \frac{mV'_y}{\sqrt{1-\frac{V'^2_x+V'^2_y}{c^2}}} \tag{3.6.23}$$

$\dot{p'}_y = p_y$ is confirmed using

$$V'_y = \sqrt{1-\frac{V'^2_x}{c^2}}\, V_y$$

$$\sqrt{1-\frac{V'^2_x+V'^2_y}{c^2}} = \sqrt{1-\frac{V_y^2}{c^2}}\sqrt{1-\frac{V'^2_x}{c^2}} \tag{3.6.24}$$

For an accelerating force, the momentum and energy increase, but the speed cannot exceed that of light. Using eq. (3.6.12), $\overrightarrow{p_1^4} + \overrightarrow{p_2^4} = \overrightarrow{p_3^4} + \overrightarrow{p_4^4}$ and $\overrightarrow{p_1^{4'}} + \overrightarrow{p_2^{4'}} = \overrightarrow{p_3^{4'}} + \overrightarrow{p_4^{4'}}$ are equivalent.

Comparing eqs. (3.6.21) and (3.6.22),

$$E^2 = (mc^2)^2 + c^2|\vec{p}|^2 \tag{3.6.25}$$

which is also applicable to bodies with $m = 0$, for example, photon. Light gives a radiation pressure when it is reflected by a mirror, showing that light has a momentum of $\frac{E_{light}}{c}$ (E_{light}: light energy) in the propagation direction (see eq. (3.4.10)), which is consistent with eq. (3.6.25). For the special theory of relativity, the relationship between momentum and energy is unified for all matter, with and without mass.

Chapter 2.4 shows the relationship between the electromagnetic vector potential and the momentum for the transformation $\vec{p} \rightarrow \vec{p} - q_e\vec{A}$ and $\rightarrow E - q_e\Phi$. Therefore, the following transformation is valid:

$$\begin{pmatrix} A'_x \\ A'_y \\ A'_z \\ \Phi'/c \end{pmatrix} = \begin{pmatrix} \frac{1}{\sqrt{1-(v/c)^2}} & 0 & 0 & \frac{-(v/c)}{\sqrt{1-(v/c)^2}} \\ 0 & 1 & 0 & 0 \\ 0 & 0 & 1 & 0 \\ \frac{-(v/c)}{\sqrt{1-(v/c)^2}} & 0 & 0 & \frac{1}{\sqrt{1-(v/c)^2}} \end{pmatrix} \begin{pmatrix} A_x \\ A_y \\ A_z \\ \Phi/c \end{pmatrix} \tag{3.6.26}$$

Maxwell's equations are also valid for another moving coordinate system. However, the electric and magnetic fields are different from those of the other coordinate systems. When $\vec{A} = 0$ and $\Phi \neq 0$, there is no magnetic field in the moving system $A'_x \neq 0$ and the magnetic field exists because the charged matter moves in another coordinate system.

The dependence of the electromagnetic field on the reference frame is considered using the following model: We consider an electric current as viewed by observer A as having a positive electric charge of density $+\rho$ moving with velocity $+V$ and a negative electric charge with density $-\rho$ moving with velocity $-V$. Although the electric field is zero, a magnetic field is induced by the current $2\rho V$. For observer B co-moving with the positive charge, the velocity of the negative charge is not $2V$, but $2V/(1 + V^2/c^2)$. Hence, the magnetic field from the perspective of observer B is smaller than that for observer A. Moreover, the unit of length for the negative charge is shorter, the density of the negative charge is higher than that of the positive charge, and the total electric charge density is $\rho\left[1 - \frac{1}{\sqrt{1-(2V/c)^2}}\right]$. Therefore, the electric field is nonzero.

Analysis based on relativistic theory is much more complicated compared to classical mechanics, mainly because motion in all three directions is not independent, as shown in eqs. (3.6.21-22). In addition, the equations of motion are generally not separable. For example, the two-body equation of motion is not separated from the motion of the center of mass and relative motion. Therefore, relativistic effects are often given as a correction to the solution of classical mechanics when these effects are small.

EXERCISE

Derive the pseudo-potential (Eq. (3.2.6)) when a charged particle (charge: q_c mass: m) is trapped by the electric field produced by the voltage generated by the voltage generated by a hyperboloid electrode.

$$\Phi = \Phi_0 \cos(\Omega t) \quad x^2 + y^2 - 2z^2 = -2z_0^2$$
$$\Phi = 0 \quad x^2 + y^2 - 2z^2 = r_0^2$$

(Answer)

$$P_{ps} = \frac{2q_c^2 \Phi_0^2 (x^2 + y^2 + 4z^2)}{m\Omega^2 (r_0^2 + 4z_0^2)^2}$$

REFERENCES

[1] P.G. Hurai, *Maxwell's equations*. IEEE Press, 2010.
[http://dx.doi.org/10.1002/9780470549919]

[2] P.M. Coulomb, *Premier memoire sur l'electrite et le magnetism*. Histoire de l'Academie Royale des Science, 1785, pp. 569-577.

[3] I.S. Grant, and W.R. Phillips, *Electromagnetism. Manchester Physics*. 2nd ed. John Wiley & Sons, 2008.ISBN978-0-471-92712-9

[4] J. Stewary, *Culculus-Early Transcendentals*. 7th ed. Books/Cole Cengage Learning, 2012, p. 1122.

[5] M.N.O. Sadik, *Elements of Electromagnetics*. 4th ed. Oxford Univ. Press, 2007, p. 386.

[6] D. Fleisch, *A Student's Guide to Maxwell's Equations*. Cambridge Univ. Press, 2008, p. 83.
[http://dx.doi.org/10.1017/CBO9780511984624.005]

[7] L.S. Brown, and G. Gabrielse, "Geonium theory: Physics of a single electron or ion in a Penning trap", *Rev. Mod. Phys.,* vol. 58, no. 1, pp. 233-311, 1986.
[http://dx.doi.org/10.1103/RevModPhys.58.233]

[8] S. Ulmer, C. Smorra, A. Mooser, K. Franke, H. Nagahama, G. Schneider, T. Higuchi, S. Van Gorp, K. Blaum, Y. Matsuda, W. Quint, J. Walz, and Y. Yamazaki, "High-precision comparison of the antiproton-to-proton charge-to-mass ratio", *Nature,* vol. 524, no. 7564, pp. 196-199, 2015.
[http://dx.doi.org/10.1038/nature14861] [PMID: 26268189]

[9] C-W. Chou, D.B. Hume, J.C.J. Koelemeij, D.J. Wineland, and T. Rosenband, "Frequency comparison of two high-accuracy Al+ optical clocks", *Phys. Rev. Lett.,* vol. 104, no. 7, p. 070802, 2010.
[http://dx.doi.org/10.1103/PhysRevLett.104.070802] [PMID: 20366869]

[10] N. Huntemann, C. Sanner, and B. Lipphardt, "Chr. Tamm, and E. Peik. Single-Ion Atomic Clock with", *Phys. Rev. Lett.,* vol. 116, p. 063001, 2016.
[http://dx.doi.org/10.1103/PhysRevLett.116.063001] [PMID: 26918984]

[11] E. V. Weisstein, "Uniqueness theorem. mathworld, wolfram.com. Retrieved 2019-11-2",

[12] EMS, 3D Electromagnetic Simulation Software | Kozo Keikaku Engineering Inc. (kke.co.jp)

[13] At a glance | Product | Murata Software Co., Ltd.

[14] B. Walker, "Optical Engineering Fundamentals. Spie Press 2008/ Fizeau H. Comptes Rendus Hebdomadaires des Seances de l'Academie des Science 1849; 29: 90/ Foucault L", *C. R. Hebd. Seances Acad. Sci.,* vol. 55, p. 501, 1862.

[15] A.A. Michelson, "The relative motion of the Earth and of the luminiferous ether", *Am. J. Sci.,* vol. 22, no. 128, pp. 120-129, 1881.
[http://dx.doi.org/10.2475/ajs.s3-22.128.120]

[16] A. Brillet, and J.L. Hall, "Improved Laser Test of Isotropy of Space", *Phys. Rev. Lett.,* vol. 42, no. 9, pp. 549-552, 1979.
[http://dx.doi.org/10.1103/PhysRevLett.42.549]

[17] A. Einstein, *Relativity-The Special and General Theory.*, 1916.
[http://dx.doi.org/10.4324/9780203198711]

[18] I. Kaminer, J. Nemirovsky, M. Rechtsman, R. Bekenstein, and M. Segev, "Seld-accelerating Dirac particles and prolonging the lifetime of relativistic fermions", *Nat. Phys.,* vol. 11, no. 3, pp. 261-267, 2015.
[http://dx.doi.org/10.1038/nphys3196]

CHAPTER 4

Fundamentals of Analysis in Quantum Mechanics

Abstract: Light has the dual characteristics of particles (photons) and electromagnetic waves. The photon has an energy of $E = h\nu$ (ν: frequency, h: Planck constant) and the momentum of $\vec{p} = h\vec{k}$ ($\vec{k}$: wavenumber and $\frac{1}{|\vec{k}|}$ is the wavelength). The photon density is proportional to the square of the amplitude of the electromagnetic waves. The fundamental aspect of quantum mechanics is that these characteristics apply to all matters. The properties of matters are described by wave functions. The probability of the existence of the matter is proportional to the square of the associated wave function. When a matter is localized in a limited region, it can only assume discrete values of energy because the wavelength of the matter wave must be an integral division of the region. The phase of the wavefunction has uncertainty on order 1/2 radians; therefore, position and momentum (time and energy) cannot be simultaneously determined. As the size of the localization area of the wavefunction becomes smaller, the minimum kinetic energy becomes larger because of the smaller wavelength (larger momentum uncertainty).

The Schroedinger equation was derived based on the idea that the relationship between the frequency and the wavenumber corresponds to that between energy and momentum given by classical mechanics, which makes it possible to obtain the wave functions of matters in the energy eigenstates. Several examples of solutions to the Schroedinger equation are introduced. The mixture between different energy eigenstates and the shift in the energy eigenvalues are induced by electromagnetic fields. The temporal change of the wave function (transition between different energy states) is also obtained using the Schroedinger equation.

Keywords: Adiabatic rapid passage, Backbody radiation, Bohr radius, Boson, Eigenfunction, Eigenvalue, Electric induced transparency (EIT), Fermion, Operator, Particle-wave duality, Photoelectronic effect, Rabi oscillation, Schroedinger equation, Stark shift, Uncertainty principle, Zeeman shift.

4.1. ESTABLISHMENT OF QUANTUM MECHANICS

As shown in chapter 3.4, the characteristics of light were determined to be identical to that of an electromagnetic wave. However, some phenomena could not be explained by considering light only as a wave. The spectrum distribution of the blackbody radiation (radiation from objects having finite temperature) has the following characteristics [1].

Masatoshi Kajita

(1) The energy density at the low frequency region is given by $\frac{8\pi\nu^2}{c^3}k_BT$, where ν is the frequency, T is the thermodynamic temperature T (see chapter 6.2), and k_B is the Boltzmann constant (defined to be 1.380649 J/K at 2019 [2]). This characteristic is explained by the pure wave characteristic of the radiation.

(2) With the high frequency region, the distribution is proportional to $\nu^3 \exp\left[-\frac{h\nu}{k_BT}\right]$. Here, h is the Planck constant (defined to be 6.62607004 × 10^{-34} J/Hz at 2019 [2]). This characteristic cannot be explained considering light only as a wave.

Planck solved this mystery based on the assumption that light energy can only assume $n_\alpha h\nu$ with the probability proportional to $\exp\left(-\frac{n_\alpha h\nu}{k_BT}\right)$ (see chapter 6.3), where $n_\alpha (\geq 0)$ is integer. The energy density is given by:

$$P_{BBR} = \frac{8\pi h\nu^3}{c^3}\frac{1}{\exp\left[\frac{h\nu}{k_BT}\right]-1}$$

$$h\nu \ll k_BT \quad P_{BBR} \approx \frac{8\pi\nu^2}{c^3}k_BT \quad \text{(influence of the energy gap is small)}$$

$$h\nu \gg k_BT \quad P_{BBR} \approx \frac{8\pi h\nu^3}{c^3}\exp\left[-\frac{h\nu}{k_BT}\right] \qquad (4.1.1)$$

Which is in good agreement with the experimental result. However, the validity of Planck's assumption was not confirmed at that time.

In addition, the emission of electrons from a material was observed when irradiated with incident light (called the photoelectronic effect). The experimental results for the photoelectric effect show that the energy of the emitted electrons is independent of the intensity of the light, although the number of emitted electrons is proportional to the intensity. Moreover, emission does not occur when ν is lower than a minimum threshold value ν_{min} and the energy of the emitted electron is proportional to the $\nu - \nu_{min}$ [3]. Einstein proposed a new concept of wave-particle duality; light has the characteristics of both waves and particles. The energy of each particle (called a photon) is $E = h\nu$ and the momentum in the propagation direction is $|\vec{p}| = h|\vec{k}| = \frac{h}{\lambda}$ (as shown in chapter 3.4), where $\vec{k}$ is the wavenumber vector and λ is the wavelength. Planck's assumption was also explained by the particle-wave duality of light. This duality was a special characteristic of light until the concept of matter waves was proposed.

There has also been a mystery regarding atomic structure since the discovery of electrons [4]. Atoms must have a structure that includes electrons, which cannot be explained by classical mechanics. Rutherford's scattering experiment [5] revealed that atoms consist of a nucleus with a positive charge and negatively charged electrons that orbit the nucleus. However, the circular motion of electrons leads to the emission of radiation energy and the corresponding loss of kinetic energy (see chapter 3.5). Electrons should crash into the nucleus after losing kinetic energy. Therefore, it was a mystery that electrons remain in their orbit. There was also another mystery in that only discrete frequency components were observed from the emission of hydrogen atoms [6].

Bohr established the "Old Quantum Mechanics" in 1913, assuming that particles bounded in a limited region of q must satisfy the following relationship [7].

$$\oint p_q dq = nh \quad n\text{: integer, } p_q\text{: momentum in the } q\text{-direction} \qquad \textbf{(4.1.2)}$$

Applying this assumption to the electron in orbit (radius r and velocity v)

$$2\pi r p = 2\pi r \mu_e v = nh \qquad \textbf{(4.1.3)}$$

where μ_e is the reduced mass between the electron and the nuclear. From the balance between the centrifugal force and the Coulomb force,

$$\frac{\mu_e v^2}{r} = \frac{e^2}{4\pi\varepsilon_0 r^2} \quad e\text{: unit electric charge} \qquad \textbf{(4.1.4)}$$

From eqs. (4.1.3-4), the possible radius of the electron orbit is given by:

$$r = a_B n^2 \quad a_B = \frac{\varepsilon_0 h^2}{\pi \mu_e e^2} \qquad \textbf{(4.1.5)}$$

where a_B is called the Bohr radius. The possible electron energy is given by:

$$E_e = \frac{\mu_e}{2} v^2 - -\frac{e^2}{4\pi\varepsilon_0 r} = -\frac{e^2}{8\pi\varepsilon_0 a_B} \frac{1}{n^2} \qquad \textbf{(4.1.6)}$$

Given that there are discrete energy levels, electrons can remain in orbits. The change in the electron energy is accompanied by the absorption or emission of photons such that we have:

$$\Delta E_e = h\nu \tag{4.1.7}$$

and the total energy is conserved. The estimated frequencies from eqs. (4.1.6-7) are in good agreement with the observed frequency components of emissions from hydrogen atoms [6]. In 1925, de Broglie presented the idea of a matter wave. He insisted that all particles have a wave-particle duality [8]. The relation $E = h\nu$ and $|\vec{p}| = h|\vec{k}| = \frac{h}{\lambda}$ is valid for all matter waves, not only for light. Equation (4.1.2) denotes that the matter waves can exist only for an integer multiple of wavelengths.

4.2. FUNDAMENTALS OF QUANTUM MECHANICS

The energy density of light is proportional to the square of the amplitudes of the electromagnetic field (eq. (3.4.7)). Although chapter 3.4 shows the oscillation of the electromagnetic field using sinusoidal functions, it is more convenient to use a complex function $\exp(ix) = e^{ix} = \cos x + i \sin x$, because the absolute value does not change with the oscillation ($|\exp(ix)| = 1$). The change in the absolute value indicates a change in amplitude. Note that the energy density is proportional to the density of the photon number. In quantum mechanics, the properties of all particles are given using wave function Ψ, and the existence probability is given by $|\Psi|^2$, in analogy with the relation between the photon density and the oscillation amplitude of the electromagnetic field. The physical value X is given as follows:

$$\int \Psi^* \check{X} \Psi d\vec{r} \tag{4.2.1}$$

as the average of the given states. Here, $\check{X}$ is the operator for X. Wave functions with deterministic values of X (called eigenvalue X_e) are called eigenfunctions Ψ_e, and they satisfy the following:

$$\check{X}\Psi_e = X_e\Psi_e \tag{4.2.2}$$

All functions can be expressed as a linear combination of eigen functions as follows:

$$\Psi = \sum a_i \Psi_{ei} \tag{4.2.3}$$

The following relation is satisfied:

$$\int \Psi_{ei}^* \Psi_{ei} d\vec{r} = 1 \quad \int \Psi_{ei}^* \Psi_{ej} d\vec{r} = 0 \ (i \neq j)$$

$$\int \Psi^* \check{X} \Psi d\vec{r} = \sum |a_i|^2 X_{ei} \qquad \sum |a_i|^2 = 1 \tag{4.2.4}$$

and the measurement result of X is one of the eigenvalues X_{ei} with a probability of $|a_i|^2$. Owing to the measurement, the wavefunction changes to the eigenfunction corresponding to the measured eigenvalue. The change in the quantum state by the measurement is called "quantum destruction".

Here, we discuss the possibility of simultaneously obtaining two deterministic physical values. The measurement uncertainties of X and Y are given by $(\check{X} - X_e)^2$ and $(\check{Y} - Y_e)^2$. For the operators, we assumed $\check{X}\check{Y} - \check{Y}\check{X} = \delta$. Thus, we have

$$[(\check{X} - X_e) - i(\check{Y} - Y_e)][(\check{X} - X_e) + i(\check{Y} - Y_e)] > 0$$

$$(\check{X} - X_e)^2 + (\check{Y} - Y_e)^2 + i(\check{X}\check{Y} - \check{Y}\check{X}) > 0$$

$$\min[(\Delta X)^2 + (\Delta Y)^2] = 2\Delta X \Delta Y > |\delta| \qquad \Delta X = |X - X_e| \quad \Delta Y = |Y - Y_e| \tag{4.2.5}$$

Then X and Y cannot be simultaneously deterministic when $\delta \neq 0$ and the relationship between the uncertainties of both values is given by $\Delta X \Delta Y > |\delta|/2$, which is called the "uncertainty principle".

For the deterministic values of energy E and momentum $\vec{p}$, the wavefunction is given by

$$\Psi = \exp\left[i\frac{2\pi}{h}(Et + \vec{p} \cdot \vec{r})\right] \tag{4.2.6}$$

and

$$H\Psi = \frac{h}{2\pi i}\frac{\partial \Psi}{\partial t} = E\Psi, \quad \frac{h}{2\pi i}\frac{\partial \Psi}{\partial q} = p_q \Psi \quad q = x, y, z \tag{4.2.7}$$

where H is the Hamiltonian operator. Note:

$$Ht - tH = p_q q - q p_q = \frac{h}{2\pi i} \tag{4.2.8}$$

Equations (4.2.5-8) show:

$$\Delta E \Delta t > \frac{h}{4\pi}, \quad \Delta p_q \Delta q > \frac{h}{4\pi} \tag{4.2.9}$$

The phase of the wavefunction given by Et/h and $p_q q/h$ has uncertainty on order of $\pm 1/2$ radians. For a deterministic p_q, the wavefunction is given by eq. (4.2.5) and $|\Psi|^2$ is completely uniform in the q-direction. There is no information regarding space distribution when the momentum is deterministic. When p_q is distributed uniformly with $-p_a < p_q < p_a$, we have:

$$\Psi = \int_{-p_a}^{p_a} \exp\left(\frac{2\pi i p_q q}{h}\right) dp_q$$

$$= \frac{h}{2\pi i q}\left[\exp\left(\frac{2\pi i p_a q}{h}\right) - \exp\left(-\frac{2\pi i p_a q}{h}\right)\right] = \frac{h}{\pi q}\sin\left(\frac{2\pi p_a q}{h}\right)$$

$$|\Psi|^2 = \left[\frac{h}{\pi}\frac{\sin\left(\frac{2\pi p_a q}{h}\right)}{q}\right]^2 \tag{4.2.10}$$

$|\Psi|^2$ is zero at $q = \pm h/2p_a$ and the particle is localized in the region $-h/2p_a < q < h/2p_a$. As the momentum distribution increases, the particle can be localized in a smaller area.

From the relationship between energy and momentum given by the classical mechanics equation:

$$E = \frac{p_x^2 + p_y^2 + p_z^2}{2m} + V(x, y, z)$$

$$V(x,y,z)\text{: potential energy} \quad m\text{: mass} \tag{4.2.11}$$

the Schroedinger equation

$$H\Psi = \frac{h}{2\pi i}\frac{\partial \Psi}{\partial t} = \left[-\frac{h^2}{8\pi^2 m}\left(\frac{\partial^2}{\partial x^2} + \frac{\partial^2}{\partial y^2} + \frac{\partial^2}{\partial z^2}\right) + V(x, y, z)\right]\Psi. \tag{4.2.12}$$

was derived. The Schroedinger equation is useful for obtaining energy eigenfunctions and eigenvalues. It is also useful for determining temporal changes in wave functions.

4.3. SOLUTION OF ENERGY EIGENVALUES

The wave function with a deterministic value of energy E is given by

$$\Psi = \exp\left(\frac{2\pi i E}{h} t\right) \varphi(\vec{r})$$
$$H\varphi = E\varphi \qquad\qquad (4.3.1)$$

Solutions of energy eigenvalues are obtained with several simple examples.

The temporal change in $|\Psi|^2$ is possible when different energy eigenstates are coupled. We consider the temporal change in the density distribution from the interference between the two components of energy and momentum as follows:

$$\Psi = \exp\left[\frac{2\pi i}{h}(E_1 t + p_1 x)\right] + \exp\left[\frac{2\pi i}{h}(E_2 t + p_2 x)\right]$$
$$\Psi^* = \exp\left[-\frac{2\pi i}{h}(E_1 t + p_1 x)\right] + \exp\left[-\frac{2\pi i}{h}(E_2 t + p_2 x)\right]$$
$$|\Psi|^2 = 2 + 2\cos\left[\frac{2\pi}{h}(E_1 - E_2)t + \frac{2\pi}{h}(p_1 - p_2)x\right] \qquad (4.3.2)$$

The motion velocity of the probability distribution is given by:

$$nv = \frac{E_1 - E_2}{p_1 - p_2} \rightarrow \vec{v} = \frac{dE}{d\vec{p}} = \frac{\vec{p}}{m} \qquad\qquad (4.3.3)$$

which corresponds to the definition of momentum in classical mechanics (eq. (2.1.2)). The velocity given by eq. (4.3.3) is called the "group velocity," which is quite different from the velocity of the phase propagation $\frac{E}{p}$. The velocity of the phase propagation is higher than the speed of light. However, it does not influence physical phenomena, and the theory of relativity is not violated. For another observer moving with a velocity of $\vec{v'}$, the observed frequency (energy) has a Doppler shift, and the group velocity is given by:

$$E + \Delta E = E\left(1 - \frac{\vec{p}\vec{v'}}{E}\right) = E - \vec{p} \cdot \vec{v'}$$
$$\frac{\vec{p} + \overrightarrow{\Delta p}}{m} = \frac{d[E + \Delta E]}{d\vec{p}} = \frac{dE}{d\vec{p}} - \vec{v'} = \vec{v} - \vec{v'} \qquad (4.3.4)$$

4.3.1. Solution for a One-dimensional Potential Well

In free space, the one-dimensional Schroedinger equation is:

$$H\varphi = -\frac{h^2}{8\pi^2 m}\frac{\partial^2}{\partial x^2}\varphi = E\varphi \tag{4.3.5}$$

and its solution is:

$$\varphi = c_+\exp\left(\frac{2\pi i\sqrt{2mE}}{h}x\right) + c_-\exp\left(-\frac{2\pi i\sqrt{2mE}}{h}x\right)$$

$$= C_p \sin\left[\frac{2\pi\sqrt{2mE}}{h}x + \delta_p\right] \tag{4.3.6}$$

Here we consider a potential well ($V(x) = 0$ at $0 < x < L$, and $V(x) = \infty$ at $x < 0$ and $x > L$). The solution for $0 < x < L$ is given by eqs. (4.3.5-6) with the requirement of $\varphi = 0$ at $x = 0, L$. Therefore,

$$\varphi \propto \sin\left[\frac{2\pi\sqrt{2mE_{n_t}}}{h}x\right] = \sin\left[\frac{n_t\pi}{L}x\right] \qquad n_t: \text{integer} \tag{4.3.7}$$

and

$$E_{n_t} = \frac{1}{2m}\left(\frac{hn_t}{2L}\right)^2 \tag{4.3.8}$$

is obtained as shown in Fig. (**4.1**).

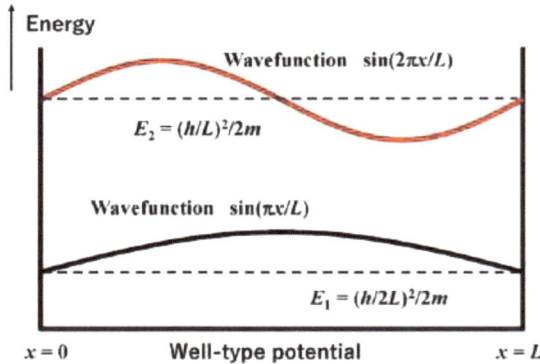

Fig. (4.1). The wave form and possible energy of atoms or molecules trapped in a potential well. (This figure is used also in "Cold Atoms and Molecules" by M. Kajita).

The non-zero lowest energy is interpreted from the uncertainty principle between momentum and position. Limiting the position region as $0 < x < L$, the uncertainty of the momentum is $\frac{h}{4\pi L}$ and the lowest energy should be larger than $\frac{1}{2m}\left(\frac{h}{4\pi L}\right)^2$. The lowest energy using the Schroedinger equation is generally larger than that obtained from the uncertainty principle. The uncertainty principle only provides the minimum broadening of the momentum as the order estimation.

The wavefunction is the standing wave without a propagation direction. This is because the wavefunction consists of the combination of momentum in the positive and negative directions.

4.3.2. Harmonic Potential

The Hamiltonian of the harmonic potential between two masses m_1 and m_2 with a vibrational frequency of ν_{vib} is given by:

$$H\varphi = \left[\frac{p^2}{2\mu_r} + \frac{\mu_a}{2}(2\pi\nu_{vib})^2 x^2\right]\varphi = E_{vib}\varphi$$
$$\mu_r = \frac{m_1 m_2}{m_1 + m_2}. \qquad (4.3.9)$$

Here we consider the operators given by:

$$a^{\pm} = \sqrt{\frac{1}{2\mu_r}}\,p \pm i\sqrt{\frac{\mu_r}{2}}(2\pi\nu_{vib})x \qquad (4.3.10)$$

It is difficult to solve eq. (4.3.9) as a second-order differential. However, using the operators $a^{\pm}$, the energy eigenvalues and eigenfunctions can be easily calculated, as shown below.

Using eq. (4.2.8), eq. (4.3.9) is rewritten as:

$$H = a^{-}a^{+} - \frac{h\nu_{vib}}{2} = a^{+}a^{-} + \frac{h\nu_{vib}}{2} \qquad a^{-}a^{+} - a^{+}a^{-} = h\nu_{vib} \qquad (4.3.11)$$

When $\Psi(\varepsilon) = \exp\left(\frac{2\pi i\varepsilon}{h}\right)\varphi(\varepsilon)$ is the eigen function with the eigenenergy of ε:

$$H[a^{\pm}\Psi(\varepsilon)] = \left[a^{\mp}a^{\pm}a^{\pm} \mp \frac{h\nu_{vib}}{2}a^{\pm}\right]\Psi(\varepsilon)$$
$$= \left[(a^{\pm}a^{\mp} \pm h\nu_{vib})a^{\pm} \mp \frac{h\nu_{vib}}{2}a^{\pm}\right]\Psi(\varepsilon)$$

$$= a^{\pm}\left(a^{\mp}a^{\pm} \pm \frac{h\nu_{vib}}{2}\right)\Psi(\varepsilon)$$

$$= a^{\pm}(H \pm h\nu_{vib})\Psi(\varepsilon)$$

$$= (\varepsilon \pm h\nu_{vib})\left(a^{\pm}\Psi(\varepsilon)\right) \tag{4.3.12}$$

and $\left(a^{\pm}\Psi(\varepsilon)\right) \propto \Psi(\varepsilon \pm h\nu_{vib})$ is derived.

We can provide another interpretation of eq. (4.3.12). From the constancy of the total energy ε, the following expression is obtained:

$$\sqrt{\frac{1}{2\mu_r}}\,p = \sqrt{\varepsilon}\cos(2\pi\nu_{vib}t) \qquad \sqrt{\frac{\mu_r}{2}}\,(2\pi\nu_{vib})x = \sqrt{\varepsilon}\sin(2\pi\nu_{vib}t) \tag{4.3.13}$$

Then

$$a^{\pm} = \sqrt{\varepsilon}[\cos(2\pi\nu_{vib}t) \pm i\sin(2\pi\nu_{vib}t)] = \sqrt{\varepsilon}\exp(\pm 2\pi i\nu_{vib}t)$$

$$\left(a^{\pm}\varphi(\varepsilon)\exp\left[\frac{2\pi i}{h}\varepsilon t\right]\right) \propto \exp\left[\frac{2\pi i}{h}(\varepsilon \pm h\nu_{vib})t\right] \to a^{\pm}\Psi(\varepsilon) \propto \Psi(\varepsilon \pm h\nu_{vib}) \tag{4.3.14}$$

To prohibit a negative value of the eigen energy, minimum energy eigenvalue of $\varepsilon_0 = \frac{h\nu_{vib}}{2}$ and $a^{-}\varphi(\varepsilon_0) = 0$ are required. $\varphi(\varepsilon_0)$ is obtained as follows:

$$a^{-}\varphi(\varepsilon_0) = 0$$

$$\frac{\partial}{\partial x}\varphi(\varepsilon_0) = -\frac{4\pi^2\mu_r\nu_{vib}}{h}x\varphi(\varepsilon_0) \qquad \int\frac{d\varphi(\varepsilon_0)}{\varphi(\varepsilon_0)} = -\frac{4\pi^2\mu_r\nu_{vib}}{h}\int x\,dx$$

$$\varphi(\varepsilon_0) \propto \exp\left(-\frac{2\pi^2\mu_r\nu_{vib}}{h}x^2\right) \quad \text{(see eq.(1.4.4))} \tag{4.3.15}$$

The higher energy eigenvalues and functions are given by:

$$\varepsilon_{n_v} = \left(n_v + \frac{1}{2}\right)h\nu_{vib}$$

$$\varphi(\varepsilon_{n_v}) = (a^{+})^{n_v}\varphi(\varepsilon_0) \text{ for example, } \varphi(\varepsilon_1) \propto x\,\mathrm{e}\,\mathrm{xp}\left(-\frac{2\pi^2\mu_r\nu_{vib}}{h}x^2\right) \tag{4.3.16}$$

There are no other energy eigenvalues. If other energy eigenvalues are possible, negative energy eigenvalues are also possible.

4.4. SCHROEDINGER EQUATION IN THREE-DIMENSIONAL POLAR COORDINATES

When the potential is given by a central force (depending only $r = \sqrt{x^2 + y^2 + z^2}$), it is convenient to use polar coordinates (r, θ, ϕ), $x = r\sin\theta\cos\phi$, $y = r\sin\theta\sin\phi$, $z = r\cos\theta$. Then eq. (4.2.12) can be rewritten as:

$$H = \left[-\frac{h^2}{8\pi^2 m}\left(\frac{\partial^2}{\partial r^2} + \frac{2}{r}\frac{\partial}{\partial r} \right) + \frac{1}{2mr^2}\widetilde{L^2} + V(r, \theta, \phi) \right] \qquad \textbf{(4.4.1)}$$

$$\widetilde{L^2} = \widetilde{L_x^2} + \widetilde{L_y^2} + \widetilde{L_z^2} = \left(\frac{h}{2\pi} \right)^2 \left[\frac{\partial^2}{\partial\theta^2} + \frac{\cos\theta}{\sin\theta}\frac{\partial}{\partial\theta} + \frac{1}{(\sin\theta)^2}\frac{\partial^2}{\partial\phi^2} \right] \qquad \textbf{(4.4.2)}$$

Where $\widetilde{L_q}(q = x, y, z)$ is the angular momentum component, as described in chapter 4.4.1. When the potential is spherically symmetric (V has no dependence on θ and ϕ), the angular momentum is constant, as shown in chapter 2.3. The wavefunction Ψ of an energy eigenstate with a spherically symmetric potential is given by the product of the functions of each parameter and the eigenvalues of the angular momentum and the energy are obtained as follows.

$$\Psi = \exp\left[\frac{2\pi i}{h}Et \right] \varphi(r, \theta, \phi)$$

$$\varphi(r, \theta, \phi) = R(r)Y(\theta)\Theta(\phi) \qquad \textbf{(4.4.3)}$$

4.4.1. Angular Momentum

First, we should consider the angular momentum, which is conserved when the potential is spherically symmetric (see chapter 2.2-3). The angular momentum operators in each direction are given as:

$$\widetilde{L_x} = yp_z - zp_y = \frac{h}{2\pi i}\left[y\frac{\partial}{\partial z} - z\frac{\partial}{\partial y} \right] = \frac{h}{2\pi i}\left[-\sin\phi\frac{\partial}{\partial\theta} - \frac{\cos\theta}{\sin\theta}\cos\phi\frac{\partial}{\partial\phi} \right]$$

$$\widetilde{L_y} = zp_x - xp_z = \frac{h}{2\pi i}\left[z\frac{\partial}{\partial x} - x\frac{\partial}{\partial z} \right] = \frac{h}{2\pi i}\left[\cos\phi\frac{\partial}{\partial\theta} - \frac{\cos\theta}{\sin\theta}\sin\phi\frac{\partial}{\partial\phi} \right]$$

$$\widetilde{L_z} = xp_y - yp_x = \frac{h}{2\pi i}\left(x\frac{\partial}{\partial y} - y\frac{\partial}{\partial x} \right) = \frac{h}{2\pi i}\frac{\partial}{\partial\phi} \qquad \textbf{(4.4.4)}$$

Equation (4.4.2) is derived from eq. (4.4.4). The eigen function Θ (satisfying $\int |\Theta(\phi)|^2 d\phi = 1$) and eigen value of L_z are given by:

$$\frac{h}{2\pi i}\frac{\partial}{\partial\phi}\Theta(\phi) = L_z\Theta(\phi)$$

$$\Theta(\phi) = \frac{1}{\sqrt{2\pi}}\exp(iM\phi) \qquad L_z = \frac{h}{2\pi}M \qquad (4.4.5)$$

Where M is the quantum number of L_z. M must be an integer given that $\Theta(\phi + 2\pi) = \Theta(\phi)$. From the uncertainty principle between ϕ and L_z, the distribution of $|\Theta|^2$ is perfectly uniform with ϕ. Therefore, $L_{x,y}$ is not deterministic except when $L_x = L_y = L_z = 0$ because we cannot determine the direction of the rotational axis in the xy-directions. The uncertainties of $L_{x,y}$ are derived from the exchange relationships between different angular momentum components shown below, which are derived using eq. (4.4.4):

$$\widetilde{L_x}\widetilde{L_y} - \widetilde{L_y}\widetilde{L_x} = i\frac{h}{2\pi}\widetilde{L_z}, \ \widetilde{L_y}\widetilde{L_z} - \widetilde{L_z}\widetilde{L_y} = i\frac{h}{2\pi}\widetilde{L_x}, \ \widetilde{L_z}\widetilde{L_x} - \widetilde{L_x}\widetilde{L_z} = i\frac{h}{2\pi}\widetilde{L_y} \qquad (4.4.6)$$

Using $\widetilde{L_\pm} = \widetilde{L_x} \pm i\widetilde{L_y}$

$$\widetilde{L_+}\widetilde{L_-} - \widetilde{L_-}\widetilde{L_+} = 2\frac{h}{2\pi}\widetilde{L_z} \quad \widetilde{L_z}\widetilde{L_+} - \widetilde{L_+}\widetilde{L_z} = \frac{h}{2\pi}\widetilde{L_+} \quad \widetilde{L_z}\widetilde{L_-} - \widetilde{L_-}\widetilde{L_z} = -\frac{h}{2\pi}\widetilde{L_-} \qquad (4.4.7)$$

As shown below, $\widetilde{L_\pm}$ makes it possible to obtain the eigenvalue and eigenfunction of the total angular momentum by solving a first-order differential function. Using eq. (4.4.7), we have:

$$\widetilde{L_z}\left(\widetilde{L_\pm}\Theta_M\right) = \left(\widetilde{L_\pm}\widetilde{L_z} \pm \frac{h}{2\pi}\widetilde{L_\pm}\right)\Theta_M = \frac{h}{2\pi}(M \pm 1)\left(\widetilde{L_\pm}\Theta_M\right)$$
$$\widetilde{L_\pm}\Theta_M \propto \Theta_{M\pm 1} \qquad (4.4.8)$$

This relationship can also be obtained from the following:

$$\widetilde{L_\pm} = \frac{h}{2\pi i}\exp(\pm i\phi)\left[\pm i\frac{\partial}{\partial\theta} - \frac{\cos\theta}{\sin\theta}\frac{\partial}{\partial\phi}\right]. \qquad (4.4.9)$$

which shows that $\widetilde{L_\pm} \propto \exp(\pm i\phi) \rightarrow \widetilde{L_\pm}\Theta_M \propto \exp[i(M \pm 1)\phi] \propto \Theta_{M\pm 1}$. We consider the eigenfunction and eigenvalue of the total angular momentum using a quantum number L as the maximum value of $|M|$. The following relationship is derived from eq. (4.4.7):

$$\widetilde{L_x}^2 + \widetilde{L_y}^2 = \widetilde{L_-}\widetilde{L_+} + \frac{h}{2\pi}\widetilde{L_z} = \widetilde{L_+}\widetilde{L_-} - \frac{h}{2\pi}\widetilde{L_z}. \qquad (4.4.10)$$

We consider the eigenfunction of the square of total angular momentum $Y_{L,M}(\theta)\Theta(\phi)$. Given that the function of $M = \pm(L + 1)$ must be forbidden, we have:

$$\widetilde{L_{\pm}}Y_{L,\pm L}(\theta)\Theta_{\pm L}(\phi) = 0 \qquad (4.4.11)$$

Then we have:

$$\widetilde{L^2}Y_{L,\pm L}(\theta)\Theta_{\pm L}(\phi) = \left[\widetilde{L_{\mp}}\widetilde{L_{\pm}} \pm \frac{h}{2\pi}\,\widetilde{L_z} + \widetilde{L_z^2}\right]Y_{L,\pm L}(\theta)\Theta_{\pm L}(\phi)$$

$$= \left(\frac{h}{2\pi}\right)^2 L(L + 1)Y_{L,\pm L}(\theta)\Theta_{\pm L}(\phi) \qquad (4.4.12)$$

The eigenvalue of the square of the angular momentum is $L(L+1)$, where L is an integer. Here eq. (4.4.8) can be rewritten using eqs. (4.4.9-12).

$$\left|\widetilde{L_{\pm}}\Theta_M\right|^2 = \Theta_M^*\left[\widetilde{L_{\mp}}\widetilde{L_{\pm}}\right]\Theta_M = \left(\frac{h}{2\pi}\right)^2 [L(L + 1) - M(M + 1)]$$

$$\widetilde{L_{\pm}}\Theta_M = \frac{h}{2\pi}\sqrt{L(L + 1) - M(M + 1)}\Theta_{M\pm 1} \qquad (4.4.13)$$

From the relationship

$$\widetilde{L_z}\left(\widetilde{L_x}^2 + \widetilde{L_y}^2 + \widetilde{L_z}^2\right) - \left(\widetilde{L_x}^2 + \widetilde{L_y}^2 + \widetilde{L_z}^2\right)\widetilde{L_z} = 0 \qquad (4.4.14)$$

the eigenvalues of $\widetilde{L_z}$ and $\widetilde{L^2}$ are simultaneously determined.

Using eq. (4.4.9), $Y_{L,\pm L}(\theta)$ is obtained as follows:

$$\left[\pm i\frac{\partial}{\partial\theta} - \frac{\cos\theta}{\sin\theta}\frac{\partial}{\partial\phi}\right]Y_{L,\pm L}(\theta)\Theta_{\pm L}(\phi) = 0 \ .$$

$$\frac{\partial}{\partial\theta}Y_{L,\pm L}(\theta) = L\frac{\cos\theta}{\sin\theta}Y_{L,\pm L}(\theta)$$

$$\int\frac{1}{Y_{L,\pm L}(\theta)}dY_{L,\pm L}(\theta) = L\int\frac{\cos\theta}{\sin\theta}d\theta = L\int\frac{1}{\sin\theta}d(\sin\theta)$$

$$\ln\left[Y_{L,\pm L}(\theta)\right] = L\ln[\sin\theta] + const$$

$$Y_{L,\pm L}(\theta)\Theta_{\pm L}(\phi) \propto [\sin\theta]^L\ \exp(\pm iL\phi) \qquad (4.4.15)$$

and the solutions of $Y_{L,M}(\theta)\Theta(\phi)$ are obtained using $\widetilde{L}_{\pm}$ as follows:

$$Y_{L,,\pm M}(\theta)\Theta_{\pm M}(\phi) \propto \left(\widetilde{L_{\mp}}\right)^{\pm(L-M)} Y_{L,,L}(\theta)\Theta_{\pm L}(\phi)$$

$$Y_{L,\pm(L-1)}(\theta)\Theta_{\pm(L-1)}(\phi) \propto \widetilde{L_{\mp}} Y_{L,\pm L}(\theta)\Theta_{\pm L}(\phi)$$

$$\propto [\sin\theta]^{L-1}\cos\theta \, \exp[\pm i(L\mp 1)\phi] \qquad (4.4.16)$$

The solutions of $Y_{L,M}(\theta)$ that satisfy $\int |Y_{L,M}(\theta)|^2 d\theta = 1$ are given by:

$$Y_{0,0}(\theta) = \sqrt{\frac{1}{2}}, \qquad Y_{1,0}(\theta) = \sqrt{\frac{3}{2}}\cos\theta, \qquad Y_{1,\pm 1}(\theta) = \frac{\sqrt{3}}{2}\sin\theta,$$

$$Y_{2,0}(\theta) = \frac{1}{2}\sqrt{\frac{5}{2}}[3(\cos\theta)^2 - 1], \; Y_{2,\pm 1}(\theta) = \frac{\sqrt{15}}{2}\sin\theta\cos\theta, \; Y_{2,\pm 2}(\theta) = \frac{\sqrt{15}}{4}(\sin\theta)^2 \qquad (4.4.17)$$

In most textbooks, the solution is expressed as:

$$Y_{L,M}(\theta) = (-1)^{|M|}\sqrt{\frac{(2L+1)(L-|M|)!}{4\pi(L+|M|)!}}\, P_L^{|M|}(\cos\theta) \qquad (4.4.18)$$

Where $P_L^{|M|}$ is the Legendre function [9].

4.4.2. Wave Function in Free Space Using the Polar Coordinate

Using the eigenvalue of $\widetilde{L}^2$ given by eq. (4.4.12), eq. (4.4.1) can be rewritten as

$$\left[\frac{\partial^2}{\partial r^2} + \frac{2}{r}\frac{\partial}{\partial r} - \frac{L(L+1)}{r^2}\right] R_L(r) = -(2\pi k)^2 R_L(r) \quad k = \frac{\sqrt{2mE}}{h} \qquad (4.4.19)$$

Equation (4.4.19) is simplified as follows using a function $\chi_L(r) = rR_L(L)$:

$$\left[\frac{\partial^2}{\partial r^2} - \frac{L(L+1)}{r^2}\right]\chi_L(r) = -(2\pi k)^2 \chi_L(r) \qquad (4.4.20)$$

With $r \to \infty$:

$$\frac{\partial^2}{\partial r^2}\chi_L(r) = -(2\pi k)^2 \chi_L(r)$$
$$\chi_L(r) \propto \sin[2\pi kr + \delta], \; \cos[2\pi kr + \delta]$$

$$R_L(r) \propto \frac{\sin[2\pi kr + \delta_0]}{r}, \frac{\cos[2\pi kr + \delta_0]}{r} \tag{4.4.21}$$

With $r \to 0$:

$$\left[\frac{\partial^2}{\partial r^2} - \frac{L(L+1)}{r^2}\right]\chi_L(r) = 0$$
$$\chi_L(r) \propto r^{L+1}, r^{-L}$$
$$R_L(r) \propto r^L, r^{-(L+1)} \tag{4.4.22}$$

The general solution of eq.(4.4.19) is given as follows:

$$R_L(r) \propto j_L[2\pi kr] \qquad j_L\text{: spherical Bessel function}$$
$$r \to 0 \quad (2\pi kr)^L \qquad r \to \infty \quad \frac{\sin\left[2\pi kr - \frac{L\pi}{2}\right]}{2\pi kr} \tag{4.4.23}$$

or

$$R_L(r) \propto n_L[2\pi kr] \qquad n_L\text{: spherical Neumann function}$$
$$r \to 0 \quad (2\pi kr)^{-(L+1)} \qquad r \to \infty \quad \frac{\cos\left[2\pi kr - \frac{L\pi}{2}\right]}{2\pi kr} \tag{4.4.24}$$

General formula of $R_L(r)$ in free space are given by:

$$R_L(r) = \frac{1}{\sqrt{1+\beta_L^2}} (2\pi k)^{\frac{3}{2}}\{j_L[2\pi kr] + \beta_L n[2\pi kr]\} \tag{4.4.25}$$

When the space is free from any potential, including $r = 0$, $\beta_L = 0$ is required to avoid divergence at $r \to 0$. (4.4.25) All wave functions in free space are given by:

$$\varphi = \sum p_L R_L(r) Y_{L0}(\theta) \qquad p_L\text{: arbitral coefficient} \tag{4.4.26}$$

Here, $M = 0$ is required since the wave function should not depend on ϕ.

For $L = 0$, an accurate solution:

$$R_0(r) \propto (2\pi k)^{\frac{3}{2}} \frac{\sin[2\pi kr]}{2\pi kr} \tag{4.4.27}$$

is obtained. For $R_L(r)$ with $L \neq 0$,

$$R_L(r) \propto r^L \qquad r \ll \frac{1}{k} \tag{4.4.28}$$

Which shows that the distribution of the wave function at a small r becomes smaller as the angular momentum increases because of the centrifugal force.

The plane wave propagating in the z-direction is given by:

$$e^{ikz} = \sum i^L 2\sqrt{\pi(2L+1)}\frac{R_L(r)}{2\pi k}Y_{L,0}(\theta) \qquad (4.4.29)$$

4.4.3. Energy State of an Electron in a Hydrogen Atom

Here we examine the energy state of an electron in a hydrogen atom considering the Coulomb potential between the electron and the nucleus. The Schroedinger equation is given by:

$$\left[-\frac{h^2}{8\pi^2\mu_e}\left(\frac{\partial^2}{\partial r^2}+\frac{2}{r}\frac{\partial}{\partial r}\right)+\frac{1}{2\mu_e r^2}\widetilde{L}^2-\frac{e^2}{4\pi\varepsilon_0 r}\right]R_L(r) = E_e R(r)$$

μ_e: reduced mass of electron-nucleus pair $\qquad$ (4.4.30)

Equation (4.4.30) is simplified as follows:

$$\left[-\frac{\partial^2}{\partial r^2}-\frac{2}{r}\frac{\partial}{\partial r}+\frac{L(L+1)}{r^2}-\frac{2}{a_B r}\right]R(r) = \zeta R(r)$$

$$\zeta = \frac{8\pi^2\mu_e}{h^2}E_e \qquad (4.4.31)$$

where a_B is the Bohr radius defined by eq. (4.1.5). Assuming the formula:

$$R(r) = G(r)\exp(-\alpha r) \qquad (4.4.32)$$

eq. (4.4.31) can be expressed as:

$$\left[-\frac{\partial^2}{\partial r^2}G(r)+2\alpha\frac{\partial}{\partial r}G(r)-\frac{2}{r}\left(\frac{\partial}{\partial r}G(r)-\alpha G(r)\right)+\frac{L(L+1)}{r^2}G(r)-\frac{2}{a_B r}G(r)-\alpha^2 G(r)\right] = \zeta G(r) \quad (4.4.33)$$

Requiring eq. (4.4.33) as an identity, the following relation is obtained:

$$-\alpha^2 = \zeta \qquad E_e = -\frac{h^2}{8\pi^2 \mu_e}\alpha^2$$

$$H_r G(r) = H_{r1} G(r) + H_{r2} G(r) = 0$$

$$H_{r1} G(r) = 2\left[\alpha \frac{\partial}{\partial r} G(r) + \frac{1}{r}\left(\alpha - \frac{1}{a_B}\right)G(r)\right]$$

$$H_{r2} G(r) = -\frac{\partial^2}{\partial r^2} G(r) - \frac{2}{r}\frac{\partial}{\partial r}G(r) + \frac{L(L+1)}{r^2}G(r) \qquad (4.4.34)$$

It is not simple to solve eq. (4.4.34). Therefore, we consider the limits of $r \to \infty$ and $r \to 0$. Note that $H_{r1}G(r)$ and $H_{r2}G(r)$ are functions of r with dimensions of one and two orders of magnitude lower than $G(r)$. For $r \to \infty$, $H_{r2}G(r)$ is negligibly small in comparison with $H_{r1}G(r)$.

$$H_r G(r) \approx H_{r1} G(r) = 0 \,,$$

$$\frac{\partial}{\partial r} G(r) = \frac{1}{r}\left(\frac{1}{a_b\alpha} - 1\right)G(r) \,,$$

$$\int \frac{1}{G(r)} = \left(\frac{1}{a_B\alpha} - 1\right)\int \frac{dr}{r} \,,$$

$$\ln G(r) = \left(\frac{1}{a_B\alpha} - 1\right)\ln r + const \,,$$

$$G(r) \propto r^{n-1} \qquad n = \frac{1}{a_B\alpha} \,,$$

$$E_e = -\frac{h^2}{8\pi^2 \mu_e}\alpha^2 = -\frac{h^2}{8\pi^2 \mu_e}\left(\frac{1}{a_B n}\right)^2 = -\frac{e^2}{8\pi\varepsilon_0 a_B}\frac{1}{n^2} \,. \qquad (4.4.35)$$

This result is consistent with that of eq. (4.1.6). The requirement of n as an integer is derived from the following. For $r \to 0$, $H_{r1}G(r)$ is negligibly small in comparison with $H_{r2}G(r)$.

$$H_r G(r) \approx H_{r2} G(r) = 0$$

$$-\frac{\partial^2}{\partial r^2} G(r) - \frac{2}{r}\frac{\partial}{\partial r}G(r) + \frac{L(L+1)}{r^2}G(r) = 0$$

$$G(r) \propto r^L, r^{-(L+1)} \qquad (4.4.36)$$

This requires that $G(r)$ does not diverge at $r \to 0$, $G(r) \propto r^L$. As shown in chapter 4.4.1, L must be an integer. Therefore, $G_{n,L}(r) = \sum_{q=L}^{n-1} c_q r^q$ is given. To satisfy

$$H_r G(r) = 0,$$

$$c_{q-1} H_{r1} r^{q-1} + c_q H_{r2} r^q = 0 \quad \text{(for the term proportional to } r^{q-2}\text{)},$$

$$c_q[-q(q+1) + L(L+1)] = \frac{2}{a_B n} c_{q-1}(n-q-1) \tag{4.4.37}$$

is required. When $c_q \neq 0$ and $q \neq L$ are simultaneously satisfied, $c_{q-1} \neq 0$. If n is not an integer, $c_q \neq 0$ for negative values of q and the wavefunction diverges at $r \to 0$. Therefore, n must be an integer laeger than L, and $c_q = 0$ with $q < L$ and $q \geq n$. The following solutions for $R_{n,L}(r)$ for $(n, L) = (1,0), (2,0)$, and $(2,1)$ satisfying $\int |R_{n,L}|^2 r^2 dr = 1$ are obtained as follows:

$$R_{1,0}(r) = 2 \left(\frac{1}{a_B}\right)^{\frac{3}{2}} \exp\left(-\frac{r}{a_B}\right)$$

$$R_{2,1}(r) = \frac{1}{2\sqrt{6}} \left(\frac{1}{a_B}\right)^{\frac{3}{2}} \left(\frac{r}{a_B}\right) \exp\left(-\frac{r}{2a_B}\right)$$

$$R_{2,0}(r) \propto \frac{1}{\sqrt{2}} \left(\frac{1}{a_B}\right)^{\frac{3}{2}} \left(1 - \frac{r}{2a_B}\right) \exp\left(-\frac{r}{2a_B}\right) \tag{4.4.38}$$

Fig. (**4.2**) shows the wavefunction shown in eq. (4.4.38). The electron quantum states are given by the quantum numbers n (principal quantum number), L (rotational quantum number), and M (magnetic quantum number). The $L = 0,1,2$ states are called the S, P, and D states, respectively. In many other textbooks, the solution of $R_{nL}(r)$ is as follows:

$$R_{nL}(r) = \left(\frac{2}{na_B}\right)^{\frac{3}{2}} \left(\frac{2r}{na_b}\right)^L \sqrt{\frac{(n-L-1)!}{2n(n+L)!}} L_{n-L-1}^{2L+1}\left(\frac{2r}{na_b}\right) \exp\left(-\frac{r}{na_B}\right) \tag{4.4.39}$$

Where L_{n-L-1}^{2L+1} is the Laguerre polynomials [10].

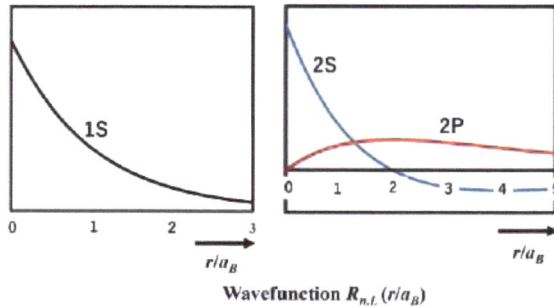

Wavefunction $R_{n,L} (r/a_B)$

Fig. (**4.2**). Wavefunction $R_{n,L}(r)$ as a functions of r/a_B (a_B: Bohr radius) for $n = 1$, $L = 0$ (1S), $n = 2$, $L = 0$ (2S), and $n = 2$, $L = 1$ (2P) states. (This figure is used also in "Cold Atoms and Molecules" by M. Kajita).

In the S state, electrons are distributed at the position of the nuclear. For a higher L state, the distribution of electrons at $r \to 0$ becomes smaller because of the centrifugal force. Although the eigenvalues of energy are the same as those obtained from the old quantum theory, the interpretation of the states is quite different. The electron motion in the S state is not a revolution but the vibration in the radial direction. The 1S state is interpreted as the electron broadening given by the uncertainty principle for p_r and r:

$$\Delta p_r \Delta r > \frac{h}{4\pi} \quad \Delta p_r \sim \frac{h}{4\pi (\Delta r)},$$ (4.4.33)

and the energy given by:

$$E_u = \frac{(\Delta p_r)^2}{2\mu_r} - \frac{e^2}{4\pi\varepsilon_0 (\Delta r)}$$ (4.4.41)

is minimum at $\Delta r = a_B/4$.

How can we provide a simple explanation for $L < n$? For a given L, the minimum effective potential energy (Coulomb potential + centrifugal force potential) is obtained when the orbital radius is as follows:

$$\frac{d}{dr}\left[\frac{L(L+1)}{r^2} - \frac{2}{a_B r}\right] = 0$$

$$r = a_B L(L+1)$$ (4.4.42)

Therefore, the electron orbital radius $a_B n^2$ should be larger than $a_B L(L+1)$.

The energy eigenvalue obtained using the Schroedinger equation depends only on n. However, the actual energy state is much more complicated because electrons have a virtual angular momentum (called spin), which can be represented as:

$$\widetilde{s^2} = \frac{3}{4}\left(\frac{h}{2\pi}\right)^2 \quad \tilde{S}_z = \frac{h}{2\pi} M_S \quad M_S = \pm\frac{1}{2}$$ (4.4.43)

The quantum number of the absolute electron spin S is 1/2, which depicts the characteristics of an electron as a permanent magnet. The two M_S states denote components in the direction of the magnet. The concept of electron spin is introduced in chapter 5 [11].

4.4.4. Energy State of Diatomic Molecules [12]

It is difficult to analyze the energy structure of molecules by considering the motion of all the nuclei and electrons. The Born-Oppenheimer approximation was proposed to simplify the analysis by focusing on the large mass ratio of atomic nuclei to electrons [13]. The electron energy state E_e is analyzed considering that the internuclear distance r is constant, as the motion of the nuclei is much slower than that of the electrons. The dependence of E_e on r yields a potential energy curve. At the electronic bonding state, E_e is a minimum for a specific value of r, called the bond length r_b, as shown in Fig. (**4.3**). E_e is generally represented by the value at $r = r_b$, because $E_e(r) - E_e(r_b)$ is given by the energy of the vibrational motion E_{vib}. The vibrational motion (around $r = r_b$) is the relative motion between the bounded nuclei and is parallel to the bonding force.

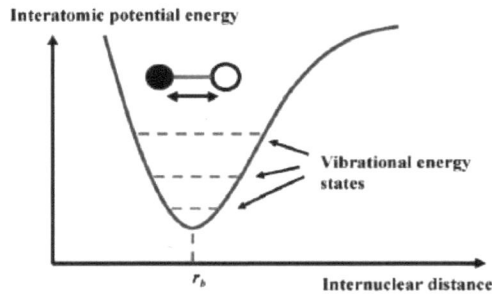

Fig. (4.3). The dependence of the electric energy state on the interatomic distance and the vibrational energy. (This figure is used also in "Cold Atoms and Molecules" by M. Kajita).

The potential energy curve was approximated by a harmonic potential at approximately $r = r_b$. As shown in chapter 4.3.2, the energy of the harmonic vibration with frequency v_{vib} is given by:

$$E_{vib} = \left(n_v + \frac{1}{2}\right) h v_{vib}$$
n_v: vibrational quantum number   (**4.4.44**)

At a higher vibrational energy state, the effect of the unharmonic potential term is significant.

The rotational motion (energy E_{rot}) is the relative motion perpendicular to the bounding direction. The rotation of molecules without electron orbital angular momentum is described by the rotational angular momentum using the quantum numbers N and M_N (integer $-N \leq M_N \leq N$), as shown in chapter 4.4.1. The angular momentum component in one direction is given by $\frac{h}{2\pi} M_N$ and the square of the absolute value of the angular momentum is $\left(\frac{h}{2\pi}\right)^2 N(N + 1)$. Then the rotational energy is given by

$$E_{rot} = hB_{n_v}N(N + 1)$$
$$B_{n_v} = \frac{h}{8\pi^2 I_{n_v}}$$
$$I_{n_v}: \text{moment inertia} \tag{4.4.45}$$

The moment of inertia is approximately represented by $I_{n_v} = \mu_a r_b^2$, but it increases slightly at higher vibrational states because the average internuclear distance is larger.

As shown in Fig. (**4.4**), the energy gap between different vibrational and rotational states is almost two and four orders of magnitude smaller than that between the electron energy states.

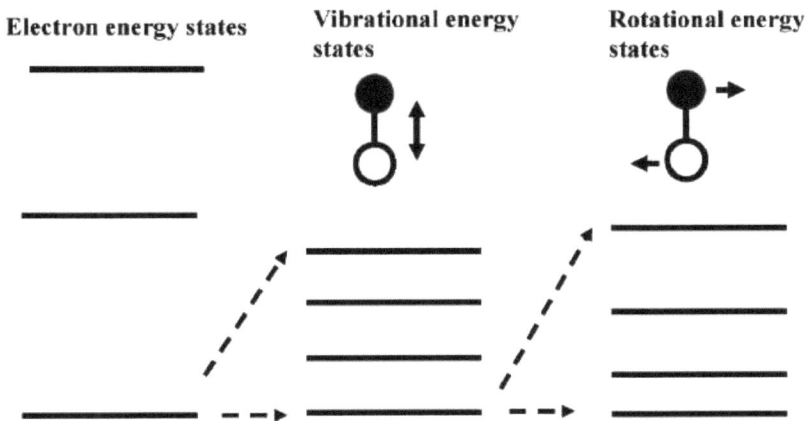

Fig. (4.4). Electron, vibrational, and rotational energy states of diatomic molecules. (This figure is used also in "Cold Atoms and Molecules" by M. Kajita).

The energy states of polyatomic molecules are much more complicated than that of diatomic molecules. In case of molecules formed by N_a atoms, there are $(3N_a - 6)$

vibrational states (for linear molecules $3N_a - 5$), and for non-linear molecules, the rotational energy depends on the direction of the rotation.

4.5. SLIGHT ENERGY SHIFT BY PERTURBATION

In this chapter, we discuss the shift in the energy eigenvalue induced by a perturbation energy term. We consider the Hamiltonian:

$$H = H_0 + H' \tag{4.5.1}$$

The energy eigenfunction is given by:

$$\varphi_i = \sum c_{ij}\, \varphi_j^0 \tag{4.5.2}$$

Where φ_j^0 is the eigenfunction of H_0, and the energy eigenvalue of E_j^0. The energy eigenvalue of H is obtained as the eigenvalues of the Hamiltonian matrix $\tilde{H}$ (chapter 1.8) with the components of:

$$
\begin{aligned}
H_{ij} &= H'_{ij} = \iiint \varphi_i^{0*} H' \varphi_j^0\, dV \quad i \neq j \\
H_{ii} &= E_i^0 + H'_{ii} = E_i^0 + \iiint \varphi_i^{0*} H' \varphi_i^0\, dV
\end{aligned} \tag{4.5.3}
$$

and c_{ij} is obtained from the eigenvector.

For simplicity, we consider two states. The Hamiltonian matrix is given by:

$$\tilde{H} = \begin{pmatrix} E_1^0 + H'_{11} & H'_{12} \\ H'_{21} & E_2^0 + H'_{22} \end{pmatrix} \tag{4.5.4}$$

and the energy eigenvalue is obtained as (see eq. (1.8.9)):

$$E^2 - (E_1^0 + E_2^0 + H'_{11} + H'_{22})E + [(E_1^0 + H'_{11})(E_2^0 + H'_{22}) - |H'_{12}|^2] = 0 \tag{4.5.5}$$

Assuming $E_1^0 \geq E_2^0$:

$$
\begin{aligned}
E_1 &= \frac{(E_1^0 + E_2^0 + H'_{11} + H'_{22}) + \sqrt{\left(E_1^0 - E_2^0 + H'_{11} - H'_{22}\right)^2 + 4|H'_{12}|^2}}{2} \\[2mm]
E_2 &= \frac{(E_1^0 + E_2^0 + H'_{11} + H'_{22}) - \sqrt{\left(E_1^0 - E_2^0 + H'_{11} - H'_{22}\right)^2 + 4|H'_{12}|^2}}{2}
\end{aligned} \tag{4.5.6}
$$

When $E_1^0 - E_2^0 + H'_{11} - H'_{22} \gg |H'_{12}| > 0$, eq. (4.5.6) is approximated to:

$$E_1 = E_1^0 + H'_{11} + \frac{|H'_{12}|^2}{E_1^0 - E_2^0 + H'_{11} - H'_{22}}$$

$$E_2 = E_2^0 + H'_{22} - \frac{|H'_{12}|^2}{E_1^0 - E_2^0 + H'_{11} - H'_{22}} \qquad (4.5.7)$$

The second and third terms on the right side denote the first- and second-order perturbation shifts, respectively.

When $H'_{12} \neq 0$,

$$E_1 - E_2 = \sqrt{(E_1^0 - E_2^0 + H'_{11} - H'_{22})^2 + 4|H'_{12}|^2} \geq 2|H'_{12}| > 0 \qquad (4.5.8)$$

and $E_1 - E_2$ cannot be negative when $H'_{11} - H'_{22} < 0$. This phenomenon is called "anti-crossing". The eigenfunctions $\varphi_{1,2}$ as a linear combination of φ_1^0 and φ_2^0 are approximately described as follow (see Fig. **4.5**):

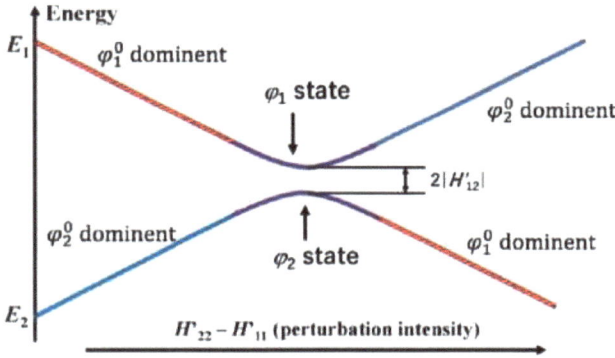

Fig. (4.5). Repulsion between two states, when the energy of the states are closed. Passing through the anti-crossing, the dominant wavefunction exchange states. Therefore, we can transform the state by changing the perturbation intensity adiabatically.

$$E_1 - E_2 \gg H'_{22} - H'_{11} \quad \varphi_1 \approx \varphi_1^0 \text{ and } \varphi_2 \approx \varphi_2^0$$

$$E_1 - E_2 = H'_{22} - H'_{11} \quad \varphi_{1,2} = \frac{\varphi_1^0 \pm \varphi_2^0}{\sqrt{2}}$$

$$E_1 - E_2 \ll H'_{22} - H'_{11} \quad \varphi_1 \approx \varphi_2^0 \text{ and } \varphi_2 \approx \varphi_1^0 \qquad (4.5.9)$$

The $\varphi_1^0 \to \varphi_2^0$ transform is possible by adiabatically changing the perturbation energy over the crossing point (see chapter 4.6.1).

For example, the perturbation can be given by the electric interaction. When voltage is present, an energy shift occurs, which is given by:

$$\Delta E_{ele} = \iiint \rho(\vec{r})\Phi(\vec{r})dV \quad \Phi: \text{voltage} \ \rho: \text{density of electric charge}$$

$$\text{taking } \Phi(\vec{r}) = \Phi(0) + \left[\frac{d\Phi(r)}{d\vec{r}}\right]_{\vec{r}=0} \vec{r} = \Phi(0) - \vec{E}(0) \cdot \vec{r}$$

$$= q_e\Phi(0) - \vec{E}(0) \cdot \overrightarrow{d_e}$$

$$q_e = \iiint \rho(\vec{r})dV \ (\text{total electric charge})$$

$$\overrightarrow{d_e} = \iiint \rho(\vec{r})\vec{r}dV (\text{electric dipole moment}) \qquad \textbf{(4.5.10)}$$

Equation (4.5.10) shows the Coulomb potential energy and Stark energy shift. The origin in eq. (4.5.10) is the center of mass. Taking the origin shifted from the center of mass as $\overrightarrow{\delta r_0}$, the dipole moment becomes $\overrightarrow{d_e} - q_e \overrightarrow{\delta r_0}$. The Coulomb potential energy is zero for neutral atoms and molecules. This energy does not induce a shift in the energy difference between different states for ions (frequency of the light absorbed or emitted) because it does not depend on the energy state. Therefore, the electric perturbation is considered mainly for the Stark energy shift. The Hamiltonian of the Stark energy shift is given by:

$$\widetilde{H_{ij}} = -\overrightarrow{d_{ij}} \cdot \vec{E} \quad \overrightarrow{d_{ij}} = \iiint \varphi_i^{0*}\rho(\vec{r})\vec{r}\varphi_j^0 dV \qquad \textbf{(4.5.11)}$$

For atoms and non-polar molecules, $\overrightarrow{d_{ii}} = 0$ because $\rho(\vec{r}) = \rho(-\vec{r})$ and the average of $\rho(\vec{r})\vec{r}$ is zero. For linear molecules, the rotational angular momentum is deterministic for the energy eigenstates, and the direction of the molecular axis is completely random. Therefore, $\overrightarrow{d_{ii}} = 0$ is also valid for linear polar molecules. The distribution of $\rho(\vec{r})$ is distorted by an electric field and the Stark energy shift is induced for $H_{ij}(i \neq j)$, which is proportional to the electric field. The energy eigenstate is obtained by the diagonalization of the Hamiltonian matrix given by eq. (4.5.11). In this state, different rotational states are mixed (angular momentum is not deterministic), and a nonzero electric dipole moment is induced in the direction parallel to the electric field. For a low electric field strength, the induced dipole moment is proportional to the electric field, and the Stark energy shift is proportional to the square of the electric field. In the case of nonlinear polar molecules, the rotational motion is a precession motion and $\overrightarrow{d_u} \neq 0$, therefore, a linear Stark energy shift occurs.

The energy shift is also induced by the magnetic field. The interaction between the magnetic dipole moments (see eqs. (2.4.6-7)) and the magnetic field induces an energy shift:

$$\Delta E_Z = \overrightarrow{\mu_m} \cdot \vec{B} \tag{4.5.12}$$

called the Zeeman energy shift. The magnetic moment is induced by the angular momentum of the electron orbital angular momentum, electron spin, nuclear spin, and molecular rotation. The angular momentum induced by the electron orbital angular momentum and molecular rotation is discussed in chapter 4.4.1. The spin of the electrons and nuclei is shown in chapter 5. This Zeeman shift is given by:

$$\Delta E_Z = \mu_B B(g_L M_L + g_S M_S + g_I M_I + g_R M_R) \quad \mu_B = \frac{eh}{4\pi m_e}$$
$$g_L = 1, \quad g_S = 2.002319, \quad g_I < 10^{-3}, \quad g_r < 10^{-4} \tag{4.5.13}$$

Where μ_B is the Bohr magneton ($\mu_B/h = 1.4$ MHz/G), g is the g-factor, and M is the quantum number of the component of the angular momentum parallel to the magnetic field. The subscripts L, S, I, and R denote the electron orbital angular momentum, electron spin, nuclear spin, and molecular rotation, respectively. From the Dirac equation, g_S is exactly 2 (see chapter 5.3). Generally, the Zeeman shift is not proportional to B, because the values of $M_{L,S,I,R}$ in eq. (4.5.13) are not constant. The wavefunction of each energy eigenstate φ_i for a given value of M is given by:

$$\varphi_i = \sum p_i^k \varphi_i^k \left(M_L^k, M_S^k, M_I^k, M_R^k \right)$$
$$M_L^k + M_S^k + M_I^k + M_R^k = M \text{ (constant)} \tag{4.5.14}$$

The splitting of the energy eigenvalues between different couplings of angular momenta (called "fine structure" and "hyperfine structure") is induced by the relativistic effect, as shown in chapter 5.4. The Hamiltonian matrix elements of the Zeeman energy are as follows:

$$\widetilde{H}_{ii} = \mu_B B \sum |p_i^k|^2 \left(g_L M_L^k + g_S M_S^k + g_I M_I^k + g_R M_R^k \right) \tag{4.5.15}$$

$$\widetilde{H}_{ij} = \mu_B B \sum p_i^{k*} p_j^k \left(g_L M_L^k + g_S M_S^k + g_I M_I^k + g_R M_R^k \right) (i \neq j) \tag{4.5.16}$$

Equation (4.5.15) gives the linear Zeeman shift considering eq. (4.5.13) with $M_{L,S,I,R} = \sum |p_i^k|^2 M_{L,S,I,R}^k$. The nondiagonal Hamiltonian matrix elements are shown in eq. (4.5.16), which induce mixture of the different energy eigenstates and the nonlinear Zeeman shift.

4.6. CHANGE OF THE ENERGY STATE WHEN THERE IS TEMPORAL CHANGE IN THE HAMILTONIAN

The uncertainty principle between time and energy shows that energy is not constant when the Hamiltonian is time-dependent. It is also useful to discuss the transition between the different energy states using the time-dependent Hamiltonian.

4.6.1. Sudden and Slow Change of the Potential

Here we assume that the initial state is an eigenstate φ_i of the Hamiltonian H_{in} (eigenvalue of energy E_i). When the Hamiltonian changes to H_{fn} rapidly (within a time shorter than $\frac{h}{E_i}$) the shape of the wavefunction cannot change. Note that an eigenfunction of H_{in} (φ_i) is given by the linear combination of eigenfunctions of H_{fn} (φ'_i) as follows:

$$\varphi_i = \sum a_{ij} \varphi'_j \tag{4.6.1}$$

By measuring the energy after the change in the Hamiltonian, $E = E'_j$ is obtained with a probability of $|a_{ij}|^2$, where a_{ij} is obtained as:

$$a_{ij} = \iiint \varphi_i^*(\vec{r}) \varphi'_j(\vec{r}) dV \tag{4.6.2}$$

For example, the wavefunction of the ground state of the harmonic potential with the frequency of ν is given by (see eq. (4.3.13)) as:

$$\varphi_0 = \left(\frac{4\pi\mu_r\nu}{h}\right)^{\frac{1}{4}} \exp\left(-\frac{2\pi^2\mu_r\nu}{h}x^2\right) \tag{4.6.3}$$

To obtain the probability of remaining in the vibrational ground state after a change in the vibrational frequency $\nu \to \nu'$, we have:

$$a_{00} = \sqrt{\frac{4\pi\mu_r\sqrt{\nu\nu'}}{h}} \int \exp\left(-\frac{2\pi^2\mu_r(\nu+\nu')}{h}x^2\right) dx$$

$$|a_{00}|^2 = \frac{2\sqrt{\nu\nu'}}{\nu+\nu'} \tag{4.6.4}$$

To analyze the case for which the Hamiltonian changes slowly, we consider the change in the vibrational frequency for N-steps. Taking $\Delta\nu = \frac{\nu'-\nu}{N}$, we consider that the vibrational frequency changes as $\nu \to \nu + \Delta\nu \to \nu + 2\Delta\nu \to \cdots\cdots \to \nu + N\Delta\nu (= \nu')$.

Considering with $\Delta\nu \ll \nu$, we have:

$$|a_{00}|^2(\nu \to \nu + \Delta\nu) = \frac{2\sqrt{\nu(\nu+\Delta\nu)}}{\nu+(\nu+\Delta\nu)} = \frac{\sqrt{1+\frac{\Delta\nu}{\nu}}}{1+\frac{\Delta\nu}{2\nu}} \approx 1 - \frac{1}{4}\left(\frac{\Delta\nu}{\nu}\right)^2 \tag{4.6.5}$$

and the probability of staying in the ground state after the change in vibrational frequency $\nu \to \nu'$ is given by:

$$|a_{00}|^2(\nu \to \nu') = \left[1 - \frac{1}{4}\left(\frac{\Delta\nu}{\nu}\right)^2\right]^N = \left[1 - \frac{1}{4}\left(\frac{\nu'-\nu}{\nu N}\right)^2\right]^N$$

$$\text{with } N \to \infty \quad |a_{00}|^2(\nu \to \nu') \to 1 \tag{4.6.6}$$

For a slow change in the Hamiltonian (adiabatic change occurring in time τ_c), the significant change in the energy is suppressed, because the uncertainty of the energy is smaller than $\frac{h}{4\pi\tau_c}$.

4.6.2. The Transition Induced by the AC Electric Field

As shown in chapter 4.5, $\vec{d} \cdot \vec{E}$ is a term of the Hamiltonian ($\vec{d}$: electric dipole moment, $\vec{E}$: electric field). Here, we discuss the transition between two states a and

b induced by the AC electric field of light in one direction $E = E_0 \cos(2\pi\nu t)$. The Hamiltonian is given by:

$$H = H_0 + H'$$

$$H_0\Psi_{a,b} = E_{a,b}\Psi_{a,b} \qquad \Psi_{a,b} = \varphi_{a,b}\exp\left(\frac{2\pi i}{h}E_{a,b}t\right)$$

$$H' = \check{d}E_0\cos(2\pi\nu t) = \check{d}E_0\frac{[\exp(2\pi i\nu t)+\exp(-2\pi i\nu t)]}{2} \qquad \textbf{(4.6.7)}$$

The temporal change of the wavefunction $\Psi = a\Psi_a + b\Psi_b$ is given by:

$$H\Psi = \frac{h}{2\pi i}\frac{\partial a}{\partial t}\Psi_a + \frac{h}{2\pi i}\frac{\partial b}{\partial t}\Psi_b + a\frac{h}{2\pi i}\frac{\partial \Psi_a}{\partial t} + b\frac{h}{2\pi i}\frac{\partial \Psi_b}{\partial t}$$
$$= aH_0\Psi_a + bH_0\Psi_b + aH'\Psi_a + bH'\Psi_b$$

$$\frac{h}{2\pi i}\frac{\partial a}{\partial t}\Psi_a + \frac{h}{2\pi i}\frac{\partial b}{\partial t}\Psi_b = aH'\Psi_a + bH'\Psi_b \qquad \textbf{(4.6.8)}$$

Here we assume:

$$d_{aa} = \iiint \varphi_a^*\check{d}\varphi_a\, dV = 0 \qquad d_{bb} = \iiint \varphi_b^*\check{d}\varphi_b\, dV = 0 \qquad \textbf{(4.6.9)}$$

Considering the product of $\Psi_{a,b}^*$ with both sides of eq. (4.6.8):

$$\frac{h}{2\pi i}\frac{\partial a}{\partial t} = bH'_{ab} \qquad H'_{ab} = d_{ab}E_0\frac{[\exp(2\pi i(\nu_0+\nu)t)+\exp(2\pi i(\nu_0-\nu)t)]}{2}$$
$$\frac{h}{2\pi i}\frac{\partial b}{\partial t} = aH'_{ba} \qquad H'_{ba} = d_{ba}E_0\frac{[\exp(-2\pi i(\nu_0+\nu)t)+\exp(-2\pi i(\nu_0-\nu)t)]}{2}$$
$$d_{ab} = \iiint \varphi_a^*\check{d}\varphi_b\, dV \qquad \nu_0 = \frac{E_b-E_a}{h} \qquad \textbf{(4.6.10)}$$

Equation (4.6.10) is simplified by ignoring the fast vibrational effect ($\propto \exp[\pm 2\pi i(\nu_0 + \nu)t]$), called rotational wave approximation. Then we have:

$$\frac{d^2b}{dt^2} = \frac{i}{\hbar}\left[2\pi i(\nu_0 - \nu)aH'_{ab} + H'_{ab}\frac{da}{at}\right] = 2\pi i(\nu_0 - \nu)\frac{db}{dt} - \frac{4\pi^2|H'_{ab}|^2}{h^2}b$$

$$\text{taking } \Delta_f = \nu_0 - \nu \quad \Omega_R = \frac{d_{ab}E_0}{h} \text{ (called Rabi frequency)}$$

$$\frac{d^2b}{dt^2} - 2\pi i\Delta_f\frac{db}{dt} + (2\pi\Omega_R)^2b \qquad \textbf{(4.6.11)}$$

As the general solution of b,

$$b = e^{i\pi\Delta_f t}\left[A \sin\left(\pi\sqrt{\Delta_f^2 + \Omega_R^2}\,t\right) + B \cos\left(\pi\sqrt{\Delta_f^2 + \Omega_R^2}\,t\right)\right] \qquad (4.6.12)$$

Assuming $b = 0$ with $t = 0$, $B = 0$, and the formula of a using A is given by:

$$a = \frac{2}{h\Omega_R e^{2\pi i\Delta_f t}}\frac{h}{2\pi i}\frac{\partial b}{\partial t} = \frac{1}{\pi\Omega_R i e^{\pi i\Delta_f t}} A\left[i\pi\Delta_f \sin\left(\pi\sqrt{\Delta_f^2 + \Omega_R^2}\,t\right) + \right.$$
$$\left. \pi\sqrt{\Delta_f^2 + \Omega_R^2}\cos\left(\pi\sqrt{\Delta_f^2 + \Omega_R^2}\,t\right)\right] \qquad (4.6.13)$$

A is obtained from the condition of $|a|^2 = 1$ with $t = 0$, and the population in the b state is given by:

$$|b|^2 = \frac{\Omega_R^2}{\Delta_f^2 + \Omega_R^2}\left(\sin\left(\pi\sqrt{\Delta_f^2 + \Omega_R^2}\,t\right)\right)^2 \qquad (4.6.14)$$

which is called the "Rabi oscillation", which is shown in Fig. (**4.6**).

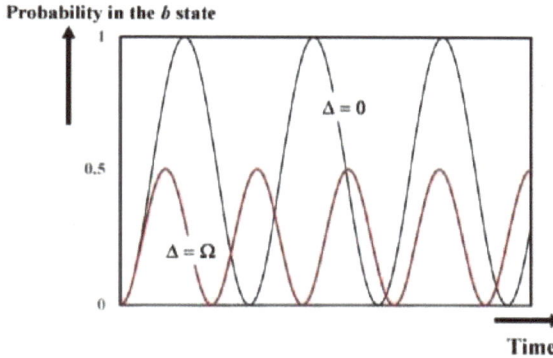

Fig. (4.6). The temporal change of the probability in the b state, when matter in the a state interacts with light with frequency detuning $\Delta = 0$ and $\Delta = \Omega$.

For a low Ω_R,

$$|b|^2 = 4\pi^2\Omega_R^2\pi t\delta\left(2\pi\Delta_f\right) \qquad (4.6.15)$$

is derived using the following formula:

$$\lim_{t\to\infty}\frac{\sin^2\left(2\pi\Delta_f t\right)}{\pi t\left(2\pi\Delta_f\right)^2} = \delta\left(2\pi\Delta_f\right) \qquad (4.6.16)$$

which is called "Fermi's golden rule".

To consider the effect of the phase jump in the wavefunction, $e^{\pm 2\pi i \Delta_f t}$ should be replaced by $e^{2\pi(\pm i \Delta_f - \gamma)t}$. Considering $e^{-2\pi \gamma t}$ as the summation of different frequency components (called the Fourier transform), we have:

$$e^{-2\pi \gamma t} = \int c_f(\delta_f) e^{2\pi i \delta_f t} d(\delta_f)$$
$$c_f(\delta_f) = \int e^{2\pi(-i\delta_f - \gamma)t} dt = \frac{1}{2\pi(i\delta_f + \gamma)}$$
$$e^{-2\pi \gamma t} = \int \frac{e^{2\pi i \delta_f t}}{2\pi(i\delta_f + \gamma)} \qquad (4.6.17)$$

Equation (4.6.14) is rewritten as:

$$|b|^2 = \frac{\Omega_R^2}{\Delta_f^2 + \gamma^2 + \Omega_R^2} \left(\sin \left(\pi \sqrt{\Delta_f^2 + \gamma^2 + \Omega_R^2} t \right) \right)^2 \qquad (4.6.18)$$

Equation (4.6.15) shows the strict conservation of the total energy, including the photon energy. However, eq. (4.6.16) shows the possible transition when $|\nu_0 - \nu| < \gamma$. This result is derived from the uncertainty principle between time and energy; the time uncertainty is $1/\gamma$ and the uncertainty of $(E_b - E_a)$ is $h\gamma$. We can determine that the phase of the wavefunction within the period of $\frac{1}{2\pi\gamma}$ is given by:

$$\phi_p = \frac{\nu_0}{\gamma} \pm 1 \qquad (4.6.19)$$

as shown in Fig. (**4.7**).

The $a \to b$ and $b \to a$ transition rates induced by light are equal. Considering only the balance of the transition rates by the incident light, the population at all energy levels is expected to be uniform. However, there is also a spontaneous emission transition from a higher energy state to a lower energy state resulting in the emission of fluorescence light, which occurs without a trigger light. From the balance between the transition rate from a lower state to a higher state (light absorption) and that from a higher state to a lower state (light induced emission + spontaneous emission transitions), the equilibrium population in the lower energy state is higher than that of, the higher energy state. When the matter is irradiated only by blackbody radiation, the energy distribution is given by the Boltzmann distribution (see chapter 6.2). The frequency of the fluorescence light is between $\nu_0 \pm \gamma$, and its phase and propagation direction are random. Spontaneous emission is a cause of the phase jump in the wave function.

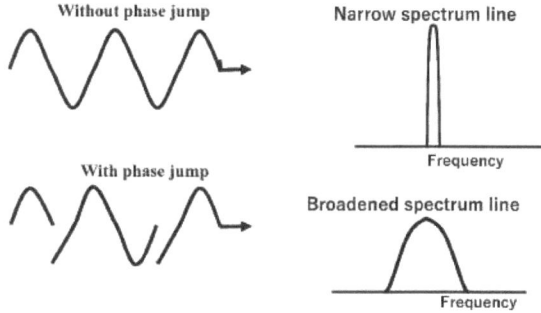

Fig. (4.7). The spectrum linewidth is broadened as the rate of the phase jump increases.

4.6.3. Electric Induced Transparency (EIT)

Chapter 4.6.2 discussed the transition between two states for a single light source. We will now examine the case of the three states a_1, a_2, and b. The energy difference between the $a_{1,2}$-b state is $h\nu_{01,2}$. When light with a frequency of $\nu_1(\nu_2)$ close to $\nu_{01}(\nu_{02})$ is considered, the a_1-b (a_2-b) transition is induced. What happens when both frequency components are irradiated simultaneously? Taking $\Psi = a_1\Psi_{a1} + a_2\Psi_{a2} + b\Psi_b$, eq. (4.6.10) can be rewritten as:

$$\frac{\partial b}{\partial t} = 2\pi i\left[\Omega_{R1}\exp\left(-2\pi i\Delta_{f1}t + i\eta_1\right)a_1 + \Omega_{R2}\exp\left(-2\pi i\Delta_{f2}t + i\eta_2\right)a_2\right]$$

$$\Delta_{f1,2} = \nu_{01,2} - \nu_{1,2}$$

$\Omega_{R1,2}$: Rabi frequency (defined in eq. (4.6.9)) between the $a_{1,2} - b$ states   **(4.6.20)**

With random $\eta_{1,2}$, both transitions are induced independently. However, when

$$\Omega_{R1}a_1 = \Omega_{R2}a_2 \qquad\qquad\qquad\qquad \textbf{(4.6.21)}$$
$$\eta_1 - \eta_2 = \pi \qquad\qquad\qquad\qquad\qquad \textbf{(4.6.22)}$$
$$\Delta_{f1} = \Delta_{f2} \qquad\qquad\qquad\qquad\qquad \textbf{(4.6.23)}$$

are satisfied, $\frac{db}{dt} = 0$, and both transitions are suppressed, as shown in Fig. (**4.8**). This phenomenon, called "Electric Induced Transparency (EIT)," is then realized.

Comparing the $a_1 \rightarrow b$ and the $a_2 \rightarrow b$ transition rates, the population ratio between the a_1 and a_2 states converges to eq. (4.6.21). Note that the phase relationship is random after spontaneous emission transition. Therefore, eq. (4.6.22) is coincidentally satisfied after repeating the cycle of the light-induced transition and the spontaneous emission transition. When eq. (4.6.23) is also satisfied; both transitions are suppressed after several cycles of the laser-induced transition and the spontaneous emission transition. After the transitions are suppressed, the states given by eqs. (4.6.21-23) are maintained.

As shown in chapter 4.6.2, the a_1-b (a_2-b) transition is induced when $|\Delta_{f1,2}| < \gamma$. Therefore, the transitions are significant when $\gamma_m < |\Delta_{f1} - \Delta_{f2}| < \gamma$ (γ_m: broadening of the transition between a_1 and a_2 states and generally $\gamma_m \ll \gamma$). EIT is used to measure the transition frequency between the a_1 and a_2 states based on the suppression of the transition to the b state when eq. (4.6.23) is satisfied [14]. EIT is also used for selective suppression of unhopeful transitions that satisfy eq. (4.6.23). For example, laser cooling is a method to reduce the atomic kinetic energy using the cycle of the laser-induced excitation and the spontaneous emission, and EIT can be used to suppress the transition that increases the kinetic energy [15, 16].

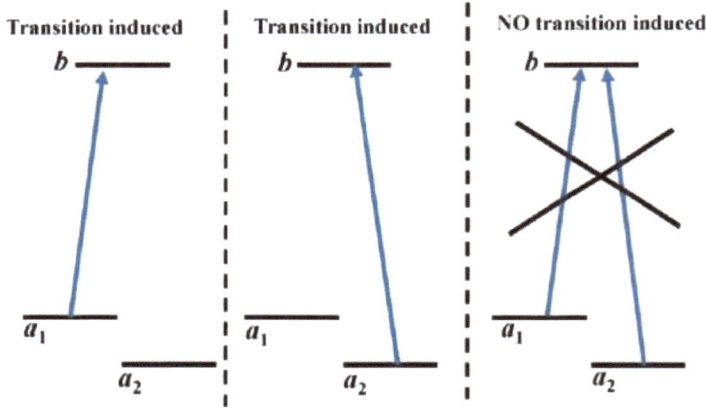

Fig. (4.8). Fundamentals of electromagnetically induced transparency (EIT). For the incident light that is tuned to the a_1–b or a_2–b transition frequency, a transition is induced. However, for the incident light with both transition frequency components, the transitions are suppressed.

As shown in eq. (4.6.21), the population ratio between a_1 and a_2 can be controlled by the intensity ratio of the transition frequency components. By reducing Ω_{R1}

adiabatically, the state population can be localized to the a_1 state. This procedure is called "stimulated Raman adiabatic passage (STIRAP)" [17].

4.6.4 State Mixture Induced by the AC Electric Field (Light)

The state mixture induced by the AC electric field (light) at a frequency of v should be discussed in terms of energy, including the photon energy. Here, we consider the Stark energy shift in the state a (energy E_a) from the mixture with state b (energy E_b). We assume $E_b - E_a = hv_0 > 0$. The total energy should be considered, including the photon energy, because the number of photons n_p changes at the transition between the two states.

We introduce a method to induce the $a \to b$ or $b \to a$ transition with a probability of 100 % using the adiabatic change of the mixing rate between the (a, n_p) and $(b, n_p - 1)$ states. Assuming $|\Delta_f| \ll v_0$ ($\Delta_f = v_0 - v$), the mixture with the $(b, n_p + 1)$ the state is negligible. The Hamiltonian matrix is expressed as follows:

$$\begin{pmatrix} E_a + n_p hv & \frac{\Omega_R}{2} \\ \frac{\Omega_R}{2} & E_b + (n_p - 1)hv \end{pmatrix} \quad \Omega_R\text{: Rabi frequency} \qquad (4.6.24)$$

and the eigen energy is given by:

$$E_\pm = h \frac{\Delta_f \pm \sqrt{\Delta_f^2 + \Omega_R^2}}{2} \qquad (4.6.25)$$

The eigenfunctions $\Psi_\pm$ corresponding to $E_\pm$ are given by:

$$\begin{aligned} \Psi_- &= \cos(\varrho)\,\Psi(a, n_p) + \sin(\varrho)\,\Psi(b, n_p - 1) \\ \Psi_+ &= -\sin(\varrho)\,\Psi(a, n_p) + \cos(\varrho)\,\Psi(b, n_p - 1) \end{aligned} \qquad (4.6.26)$$

The value of ϱ corresponds to

$$\varrho \to 0 \quad \Delta_f > 0 \quad \frac{\Omega_R}{\Delta_f} \to 0 \quad \Psi_- = \Psi(a, n_p) \quad \Psi_+ = \Psi(b, n_p - 1)$$

$$\varrho = \frac{\pi}{4} \quad \Delta_f = 0 \qquad\qquad \Psi_\pm = \frac{\Psi(a, n_p) \pm \Psi(b, n_p - 1)}{\sqrt{2}}$$

$$\varrho \to \frac{\pi}{2} \quad \Delta_f < 0 \quad \frac{\Omega_R}{\Delta_f} \to 0 \quad \Psi_- = \Psi(b, n_p - 1) \quad \Psi_+ = \Psi(a, n_p) \qquad \text{(4.6.27)}$$

Adiabatic change of ϱ from 0 to $\frac{\pi}{2}$ can be performed as follows:

(1) Initially take large positive Δ_f (large red detuning) and small Ω_R.
(2) Decrease Δ_f until $\Delta_f = 0$. Ω_R is increased so that $E_+ - E_-$ at $\Delta_f = 0$ is large
enough to avoid the jump between Ψ_- and Ψ_+ ($E_+ - E_- \geq \Omega_R \gg \frac{h}{4\pi\tau_p}$, where
τ_p is the time required for the procedure).
(3) Increase Δ_f in the negative direction (blue detuning) and Ω_R is reduced.

Using this procedure, the $a \to b$ or $b \to a$ transition is performed with the probability of 100 %. We can obtain a mixture of 50 % of the a and b states by turning off the light after the procedure (2). This procedure is called the "Adiabatic Rapid Passage [18]".

The Stark energy shift in the a-state is obtained from the difference between $E_a + n_p h\nu$ and the eigenvalue of the Hamiltonian matrix:

$$\begin{pmatrix} E_a + n_p h\nu & \frac{\Omega_R}{2} & \frac{\Omega_R}{2} \\ \frac{\Omega_R}{2} & E_b + (n_p - 1)h\nu & 0 \\ \frac{\Omega_R}{2} & 0 & E_b + (n_p + 1)h\nu \end{pmatrix} \qquad \text{(4.6.28)}$$

When $\Omega_R \ll |\Delta_f|$, the Stark energy state in the a-state is given by:

$$\Delta E_{Sa} = -\frac{h\Omega_R^2 \nu_0}{2(\nu_0^2 - \nu^2)} \qquad \text{(4.6.29)}$$

which shows that the Stark energy shift is negative for $\nu < \nu_0$ (red detuning). For $\nu > \nu_0$ (blue detuning), the Stark shift is positive. Equation (4.6.29) was approximated as follows:

$$\nu \ll \nu_0 \qquad \Delta E_{Sa} = -\frac{h\Omega_R^2}{2\nu_0} \quad \text{(DC Stark shift with the average electric field)}$$

$$\nu \approx \nu_0 \qquad \Delta E_{Sa} = -\frac{h\Omega_R^2}{4\Delta_f} \quad \text{(corresponds to eq. (4.6.22))} \qquad \text{(4.6.30)}$$

4.7. QUANTUM TREATMENT OF ELASTIC COLLISION BY A SPHERICALLY SYMMETRIC POTENTIAL

When there is a potential between atoms or molecules, collisional interactions occur when these particles get close to each other. The collision rate is given by:

$$\Gamma = n_m v_r \sigma_c \tag{4.7.1}$$

Where n_m is the density of gaseous atoms or molecules, v_r is the mean relative velocity, and σ_c is the collision cross-section. Here, we analyze the collision cross-section assuming a spherically symmetric potential. We considered the scattering of the incident plane wave (see eq. (4.4.29)) as follows:

$$e^{ikz} = \sum i^L 2\sqrt{\pi(2L+1)} \frac{R_L(r)}{2\pi k} Y_{L,0}(\theta)$$
$$R_L(r) = (2\pi k)^{\frac{3}{2}} j_L(2\pi kr) \quad k = \frac{\sqrt{2\mu_r E}}{h}$$
$$(E\text{: collision energy, } \mu_r\text{: reduced mass}) \tag{4.7.2}$$

The cross-section of each partial wave with a given L is given as:

$$\sigma_L = \frac{2L+1}{\pi k^2} \tag{4.7.3}$$

and the collisional cross-section is given as:

$$\sigma_c = \sum \sigma_L P_L = \frac{1}{\pi k^2} \sum (2L+1) P_L \tag{4.7.4}$$

where P_L is the probability of collisional scattering. P_L is obtained from the distortion of the wavefunction by the potential. Assuming a spherically symmetric potential, the dependence of the wavefunction on θ does not change with the collision. For the spherically symmetric potential at $r < a_p$, the region at $r > a_p$ is treated as the free space, and eq. (4.7.2) is corrected as follows (see eq. (4.4.24)):

$$R_L(r) = \frac{(2\pi k)^{\frac{3}{2}}}{\sqrt{1+\beta_L^2}} [j_L(2\pi kr) + \beta_L n_L(2\pi kr)] \tag{4.7.5}$$

Equation (4.7.5) is applied only for $r > a_p$, therefore, it is not required to avoid the divergence at $r \to 0$, and the spherical Neumann function can be mixed. Comparing eqs. (4.7.2) and (4.7.5),

$$P_L = \frac{\beta_L^2}{1+\beta_L^2} \tag{4.7.6}$$

For a large r,

$$j_L(2\pi kr) = \frac{\sin\left(2\pi kr - \frac{L\pi}{2}\right)}{2\pi kr} \qquad n_L(2\pi kr) = \frac{\cos\left(2\pi kr - \frac{L\pi}{2}\right)}{2\pi kr} \tag{4.7.7}$$

and

$$\frac{1}{\sqrt{1+\beta_L^2}}[j_L(2\pi kr) + \beta_L n_L(2\pi kr)] = \frac{\sin\left(2\pi kr - \frac{L\pi}{2} + \delta_L\right)}{2\pi kr}$$

$$\beta_L = \tan(\delta_L)$$

$$P_L = [\sin(\delta_L)]^2$$

$$\sigma_c = \frac{1}{\pi k^2}\Sigma(2L + 1)[\sin(\delta_L)]^2 \tag{4.7.8}$$

As an example, we consider a case of a potential well ($V(r) = 0$ and $r > a_p$ and $V(r) = \infty$ for $r < a_p$). Equation (4.7.5) is valid for $r > a_p$, and $R_L(r) = 0$ with $r < a_p$. Therefore,

$$j_L(2\pi ka_p) + \beta_L n_L(2\pi ka_p) = 0 \tag{4.7.9}$$

is required. Assuming $ka_p \ll 1$ (satisfied for atoms with kinetic energy lower than 1 mK),

$$j_L(2\pi ka_p) = (2\pi ka_p)^L \qquad n_L(2\pi ka_p) = (2\pi ka_p)^{-(L+1)} \tag{4.7.10}$$

and eq. (4.7.9) is satisfied for

$$\beta_L = (2\pi ka_p)^{2L+1} \tag{4.7.11}$$

And

$$\sigma_c = \frac{1}{\pi k^2}\sum(2L+1)\frac{(2\pi k a_p)^{4L+2}}{1+(2\pi k a_p)^{4L+2}} \qquad (4.7.12)$$

For the case of ultra-low kinetic energy, the collision cross-section is dominated by the term with $L = 0$, and it converges to $\sigma_c = 4\pi a_p^2$. This is four times larger than the term obtained from classical mechanics $\sigma_c = \pi a_p^2$. For $L \geq 1$, the distribution of the wavefunction at $r < a_p$ is negligibly small because of the centrifugal force potential.

When the interaction is weak, eq. (4.7.8) is approximated as:

$$\sigma_c = \frac{1}{\pi k^2}\sum(2L+1)\delta_L^2$$
$$\delta_L = (\Delta\tau)\frac{2\pi}{h}(2\pi k)^3 \int[j_L(2\pi k r)]^2 V(r) r^2 dr \qquad (4.7.13)$$

where $\Delta\tau$ is the interaction time given by:

$$\Delta\tau = \frac{\left(\frac{2\pi}{k}\right)}{\left(\frac{hk}{\mu_r}\right)} = \frac{2\pi\mu_r}{hk^2} \qquad (4.7.14)$$

This treatment is called the "Born approximation". For the interaction between electric dipole moments, $V(r) \propto r^{-3}$ and $\delta_L \propto k$. Therefore, $\sigma_L P_L$ is independent of the collisional kinetic energy, whereas the Born approximation is valid.

The quantum characteristic results from the interference between undistinguishable phenomena. For the collision between the identical particles in the same quantum state, there must be an interference of the relative positions $\vec{r}$ and $-\vec{r}$. For the collision between the same species, the wavefunction should be exchanged as follows:

$$\varphi_L(\vec{r}) \rightarrow \frac{\varphi_L(\vec{r})+\varphi_L(-\vec{r})}{\sqrt{2}} \quad \text{for Boson particles (with integer spin)}$$

$$\varphi_L(\vec{r}) \rightarrow \frac{\varphi_L(\vec{r})-\varphi_L(-\vec{r})}{\sqrt{2}} \quad \text{for Fermion particles (with half integer spin)} \qquad (4.7.15)$$

As $\varphi_L(-\vec{r}) = \varphi_L(\vec{r})$ for even L and $\varphi_L(-\vec{r}) = -\varphi_L(\vec{r})$ for odd L, eq. (4.7.8) is replaced as:

$$\sigma_c = \frac{2}{\pi k^2}\sum_{L=even}(2L+1)[\sin(\delta_L)]^2 \quad \text{for Boson particles}$$

$$\sigma_c = \frac{2}{\pi k^2} \Sigma_{L=odd} (2L+1)[\sin(\delta_L)]^2 \text{ for Fermion particles} \qquad (4.7.16)$$

For the collision between the same species of Fermion particles, the term $L = 0$ does not exist, and the collision effect induced by the short-range interaction is suppressed for the ultra-low kinetic energy. This effect has been experimentally observed for ^{83}K atoms [19]. This suppression is not observed when the interaction is caused by the electric dipole-dipole interaction because the collision term for $L \geq 1$ is also not suppressed for $k \to 0$ [20].

For the case of high kinetic energy, $1/k$ is small and taking

$$b_c = \frac{L}{\pi k} \quad db_c = \frac{1}{\pi k} \qquad (4.7.17)$$

eq. (4.7.8) can be rewritten as

$$\sigma_c = \int 2\pi b_c \left[\sin\big(\delta(b_c)\big) \right]^2 db \qquad (4.7.18)$$

Where b_c is the closest distance between interacting particles (called impact parameters).

EXERCISE

(1) For a potential well, the eigenfunction is given by
$$\Phi_n = \exp\left(i \frac{E_n}{h} t \right) \sin\left(\frac{nx}{L} \right)$$
Derive the temporal change of the density distribution of the matter when
$\Phi = \Phi_1 + \Phi_2$

(Answer)
$$|\Phi|^2 = \left[\sin\left(\frac{x}{L} \right) \right]^2 + \left[\sin\left(\frac{2x}{L} \right) \right]^2 + 2 \cos\left(\frac{E_2 - E_1}{h} t \right) \sin\left(\frac{x}{L} \right) \sin\left(\frac{2x}{L} \right) x$$

(2) Estimate the minimum energy of matter in a harmonic potential (frequency v) using the uncertainty principle for position x and momentum p

(Answer)
$$\frac{(\Delta p)^2}{2m} + \frac{m(2\pi v)^2 (\Delta x)^2}{2} \geq (2\pi v)(\Delta p \Delta x) \geq \frac{hv}{2}$$

REFERENCES

[1]   R. London, *The quantum theory of light.* 3rd ed. Cambridge Univ. Press, 1973.

[2]   CODATA Value: Planck constant (nist.gov)

[3]   A.T. Fromhold, *Quantum Mechanics for Applied Physics and Engineering.* Courier Dover Publications, 1991, pp. 5-6.Photoelectric effect - Wikipedia

[4]   C. John, M. Paul, and T. John, "Key experiments: how do we know the nature of the atom a nd its components?. chemistry and chemical reactivity", *Instructor's Edition, Stamford CT. Cengrade Learning,* pp. 54-55, 2015.

[5]   E. Rutherford, "The scattering of a and b particles by matter and the structure of the atom", *Philosophical Magazine,* vol. 21, pp. 669-688, 1911.

[6]   J.J. Balmer, "Notiz ueber die Spektrallinien des Wasserstoffs", *1885 Annalen der Physik und Chemie,* vol. 25, pp. 80-87, 1885. (in German).

[7]   N. Bohr, "On the constitution of atoms and molecules", *Phil. Mag. Series,* vol. 26, no. 6, pp. 1-25, 1913.

[8]   L.V. de Broglie, *"Recherches sur la theorie des quanta",* Thesis Paris 1924 Anne de Physique, vol. 10, no. 3, pp. 22-1281925.

[9]   Spherical harmonics - Wikipedia

[10]   C.F. Dunkl, "A laguerre polynominal orthogonality and the hydrogen atom", *Anal. Appl.,* vol. 1, pp. 177-185, 2003.

[11]   P.W. Atkins, *A handbook of concepts.* Oxford University Press, 1974, p. 52.

[12]   M. Kajita, *Cold atoms and molecules.* IOP Expanding Physics, 2020, pp. 10-14.

[13]   M. Born, and J.R. Oppenheimer, "Zur quantentheorie der molekeln", *Ann. Phys.,* vol. 389, pp. 457-484, 1927. [in German].

[14]   S. Knappe, V. Gerginov, P.D. Schwindt, V. Shah, H.G. Robinson, L. Hollberg, and J. Kitching, "Atomic vapor cells for chip-scale atomic clocks with improved long-term frequency stability", *Opt. Lett.,* vol. 30, no. 18, pp. 2351-2353, 2005.
        [PMID: 16196316]

[15]   A. Aspect, E. Arimondo, R. Kaiser, N. Vansteenkiste, and C. Cohen-Tannoudji, "Laser cooling below the one-photon recoil by velocity-selective coherent population trapping", *Phys. Rev. Lett.,* vol. 61, no. 7, pp. 826-829, 1988.
        [PMID: 10039440]

[16]   G. Morigi, J. Eschner, and C.H. Keitel, "Ground state laser cooling using electromagnetically induced transparency", *Phys. Rev. Lett.,* vol. 85, no. 21, pp. 4458-4461, 2000.
        [PMID: 11082570]

[17]   K. Aikawa, D. Akamatsu, M. Hayashi, K. Oasa, J. Kobayashi, P. Naidon, T. Kishimoto, M. Ueda, and S. Inouye, "Coherent transfer of photoassociated molecules into the rovibrational ground state", *Phys. Rev. Lett.,* vol. 105, no. 20, 2010.203001
        [PMID: 21231225]

[18]   V.S. Malinovski, and J.L. Klause, "General theory of population transfer with adiabatic rapid passage with intense chirped laser pulses", *Eur. Phys. J. D,* vol. 14, pp. 147-155, 2001.

[19]   H. Katori, H. Kunugita, and T. Ido, "Quantum statistical effect on ionizing collisions of ultracold metastable Kr isotopes", *Phys. Rev. A,* vol. 52, no. 6, pp. R4324-R4327, 1995.
        [PMID: 9912850]

[20]   M. Kajita, "Cold collisions between boson or fermion molecules", *Phys. Rev. A,* vol. 69, 2004.012709

CHAPTER 5

Relativistic Quantum Mechanics and Spin

Abstract: Spin is one of the most important characteristics of particles, which results in a fine and hyperfine energy structure and Zeeman energy shift. Spin is a property of particles, independent of the density distribution given by the potential field. The spin eigenfunction should be described using a vector, and the spin operator is given by a matrix. The property of spin is not derived from the Schroedinger equation, which treats wave functions as scalars.

The Dirac equation was developed to obtain wave functions from the relationship between frequency (energy) and wavenumber (momentum) given by the theory of special relativity, for which the equation is given by the 4×4 matrix and the wave functions are given by four-dimensional vectors. There are four solutions for one equation, which correspond to $\pm \frac{1}{2}$ spin states and the positive and negative rest energies. The existence of negative rest energy was confirmed by the discovery of positrons (antiparticles). The Zeeman shift induced by spin is derived from the exchange of matrix production.

Keywords: Antiparticle, Dirac equation, Electron sea, Electron spin, Fine structure, Klein-Gordon equation, Lamb shift, Pauli matrix, Quantum electrodynamics, Relativistic effect, Zeeman shift.

5.1. ELECTRON SPIN

As shown in chapter 4, the quantum energy state of an electron in an atom is derived from the Schroedinger equation using the principal quantum number n, rotational quantum number L, and magnetic quantum number M. These quantum numbers are given by the motion of electrons under a Coulomb potential owing to the nucleus. An electron is a Fermion, and only one electron can be in a quantum state (chapter 6.6). However, two electrons can be in an (n, L, M) state, which means there are two states with electrons. In the Stern-Gerlach experiment, an Ag atomic beam traversed through an area with an inhomogeneous magnetic field. The atoms were deflected from the straight path in two opposite directions. The deflection angle was the only quantized parameter [1]. The electron orbital angular momentum of the Ag atom was zero. This result shows that the electron has two states like a permanent magnet: the S-pole or N-pole in the direction of the magnetic field (see chapters 4.4 and 4.5). In analogy with the M states defined as $-L \leq M \leq L$ for each L state (number of states $2L + 1$), these two states of electrons were described as the components of a virtual angular momentum (called spin) $S = \frac{1}{2}$ in one direction,

$M_S = \pm\frac{1}{2}$. The commen characteristics of the spin and orbital angular momentum are confirmed in more detail later. The spin state is a property of the electron itself, and the eigenfunction is not described by the wave function associated with the density distribution. Pauli proposed describing two spin states using a two-dimensional vector [2]. The eigenfunction of each spin state is as follows:

$$M_S = \frac{1}{2} \;\rightarrow\; \xi_+ = \begin{pmatrix} 1 \\ 0 \end{pmatrix} \quad M_S = -\frac{1}{2} \;\rightarrow\; \xi_- = \begin{pmatrix} 0 \\ 1 \end{pmatrix} \tag{5.1.1}$$

and the general wave functions as the combination of both spin states are given by

$$\Psi = (a\xi_+ + b\xi_-) \int c(E,\vec{p}) \exp\left[\frac{2\pi i}{h}(Et + \vec{p}\cdot\vec{r})\right] dE d\vec{p}$$

$$|a|^2 + |b|^2 = 1 \tag{5.1.2}$$

The operator of the spin components in the x,y,z-directions are

$$\widetilde{S_q} = \frac{h}{4\pi}\sigma_q \quad q = x, y, z$$

$$\sigma_x = \begin{pmatrix} 0 & 1 \\ 1 & 0 \end{pmatrix}, \sigma_y = \begin{pmatrix} 0 & -i \\ i & 0 \end{pmatrix}, \sigma_z = \begin{pmatrix} 1 & 0 \\ 0 & -1 \end{pmatrix} \tag{5.1.3}$$

where σ_q are the Pauli matrices. For the Pauli matrices,

$$\sigma_x^2 = \sigma_y^2 = \sigma_z^2 = I \quad I = \begin{pmatrix} 1 & 0 \\ 0 & 1 \end{pmatrix}$$

$$\sigma_x\sigma_y + \sigma_y\sigma_x = \sigma_x\sigma_z + \sigma_z\sigma_x = \sigma_y\sigma_z + \sigma_z\sigma_y = 0$$

$$\sigma_x\sigma_y - \sigma_y\sigma_x = 2i\sigma_z, \quad \sigma_y\sigma_z - \sigma_z\sigma_y = 2i\sigma_x, \quad \sigma_z\sigma_x - \sigma_x\sigma_z = 2i\sigma_y \tag{5.1.4}$$

are satisfied. The exchange law:

$$\widetilde{S_x}\widetilde{S_y} - \widetilde{S_y}\widetilde{S_x} = \frac{h}{2\pi}i\widetilde{S_z}, \quad \widetilde{S_y}\widetilde{S_z} - \widetilde{S_z}\widetilde{S_y} = \frac{h}{2\pi}i\widetilde{S_x}, \quad \widetilde{S_z}\widetilde{S_x} - \widetilde{S_x}\widetilde{S_z} = \frac{h}{2\pi}i\widetilde{S_y} \tag{5.1.5}$$

is the same as that for the orbital angular momentum, as shown in eq. (4.4.6). The square of the absolute value is given by:

$$\left(\widetilde{S_x}\right)^2 + \left(\widetilde{S_y}\right)^2 + \left(\widetilde{S_z}\right)^2 = \left(\frac{h}{2\pi}\right)^2 \frac{3}{4} I = \left(\frac{h}{2\pi}\right)^2 S(S+1)I \tag{5.1.6}$$

Which corresponds to the square of the absolute value of the orbital angular momentum is given by $\left(\frac{h}{2\pi}\right)^2 L(L+1)$ as shown by eq. (4.4.12).

Using the operators

$$\widetilde{S_+} = \widetilde{S_x} + i\widetilde{S_y} = \frac{h}{2\pi}\begin{pmatrix} 0 & 1 \\ 0 & 0 \end{pmatrix} \quad \widetilde{S_-} = \widetilde{S_x} - i\widetilde{S_y} = \frac{h}{2\pi}\begin{pmatrix} 0 & 0 \\ 1 & 0 \end{pmatrix} \tag{5.1.7}$$

the following relations hold

$$\widetilde{S_+}\xi_+ = 0, \quad \widetilde{S_+}\xi_- = \frac{h}{2\pi}\xi_+, \quad \widetilde{S_-}\xi_+ = \frac{h}{2\pi}\xi_-, \quad \widetilde{S_-}\xi_- = 0 \tag{5.1.8}$$

which corresponds to the relationship with the orbital angular momentum as follows (see eq. (4.4.13)):

$$\widetilde{L_\pm}\Theta_M = \frac{h}{2\pi}\sqrt{L(L+1) - M(M \pm 1)}\Theta_{M\pm 1} \tag{5.1.9}$$

Therefore, the electron spin satisfies all the relations with the angular momentum, except that the quantum numbers are given as half integers.

As shown in eq. (4.5.13), the Zeeman energy shift induced by the electron spin and the magnetic field B is obtained as follows:

$$E_Z = g_S\mu_B M_S B \qquad \mu_B = \frac{eh}{4\pi m_e}$$

$g_S = 2.002319$ μ_B: Bohr magneton g_S: g-factor of the electron spin
e: unit charge m_e: electron mass   (5.1.10)

Electron spin is not derived from the Schroedinger equation treatment of the wave function as a scalar, although the Zeeman energy shift is treated as an additional perturbation term. Given that the operators of electron spin are matrices, the Hamiltonian should be a matrix. The Zeeman energy shift induced by the electron spin is derived in chapter 5.3, taking $g_S = 2$.

5.2. KLEIN-GORDON EQUATION [3]

The Schroedinger equation is based on the relationship between the frequency (energy) and wavenumber (momentum) given in classical mechanics. Based on the theory of special relativity, the relationship between energy and momentum is given by (see eq. (3.6.26)):

$$E^2 = (mc^2)^2 + c^2|\vec{p}|^2 \tag{5.2.1}$$

Using the energy and momentum operators shown in eq. (4.2.6), the Klein-Gordon equation is given as:

$$-\left(\frac{h}{2\pi}\right)^2 \frac{\partial^2}{\partial t^2} = -\left(\frac{h}{2\pi}\right)^2 c^2 \left[\frac{\partial^2}{\partial x^2} + \frac{\partial^2}{\partial y^2} + \frac{\partial^2}{\partial z^2}\right] + (mc^2)^2 \tag{5.2.2}$$

is derived. There are challenges associated with the general application of this equation because it derives the square of the energy (not energy itself), and the utility of the solutions is questionable. The main problem is that negative probability and energy values are included in the solutions. The wave function is treated as a scalar; therefore, the effect of electron spin is not included.

The Klein-Gordon equation was forgotten for a while. However, Pauli and Weisskopf determined that this equation is useful for treating Boson particles with zero spin, for which the wave function is treated as a scalar [4]. Later, Dirac interpreted negative probability and energy [5].

5.3. DIRAC EQUATION AND DERIVATION OF ELECTRON SPIN [6]

Dirac proposed an equation for a Hamiltonian as a first-order differential equation as follows:

$$\tilde{H} = c\left[\widetilde{\alpha_x}\widetilde{p_x} + \widetilde{\alpha_y}\widetilde{p_y} + \widetilde{\alpha_z}\widetilde{p_z}\right] + \tilde{\beta}mc^2 \tag{5.3.1}$$

with the requirement that $\tilde{H}^2$ corresponds to the Klein-Gordon equation. For operators $\widetilde{\alpha_q}$ ($q = x, y, z$) and $\tilde{\beta}$, the following relations must hold:

$$\widetilde{\alpha_q}^2 = \widetilde{\beta}^2 = 1$$
$$\widetilde{\alpha_x}\widetilde{\alpha_y} + \widetilde{\alpha_y}\widetilde{\alpha_x} = \widetilde{\alpha_y}\widetilde{\alpha_z} + \widetilde{\alpha_z}\widetilde{\alpha_y} = \widetilde{\alpha_z}\widetilde{\alpha_x} + \widetilde{\alpha_x}\widetilde{\alpha_z} = 0$$
$$\widetilde{\alpha_q}\widetilde{\beta} + \widetilde{\beta}\widetilde{\alpha_q} = 0 \tag{5.3.2}$$

Equation (5.3.2) cannot be satisfied for a scalar but can be satisfied using 4×4 matrices:

$$\alpha_q = \begin{pmatrix} 0 & \sigma_q \\ \sigma_q & 0 \end{pmatrix}, \beta = \begin{pmatrix} I & 0 \\ 0 & -I \end{pmatrix} \tag{5.3.3}$$

where σ_q are the Pauli matrices (see eq. (5.1.3)). The eigenfunctions are given by four-dimensional vectors

$$\vec{\Psi} = \begin{pmatrix} \vec{u} \\ \vec{w} \end{pmatrix} \quad \vec{u} = \begin{pmatrix} u_1 \\ u_2 \end{pmatrix} \quad \vec{w} = \begin{pmatrix} w_1 \\ w_2 \end{pmatrix} \tag{5.3.4}$$

At first, we consider $\vec{p} = 0$ for simplicity. Then

$$\widetilde{H} = \begin{pmatrix} mc^2 & 0 & 0 & 0 \\ 0 & mc^2 & 0 & 0 \\ 0 & 0 & -mc^2 & 0 \\ 0 & 0 & 0 & -mc^2 \end{pmatrix} \tag{5.3.5}$$

and four eigenfunctions

$$\vec{\Psi} = \begin{pmatrix} 1 \\ 0 \\ 0 \\ 0 \end{pmatrix}, \begin{pmatrix} 0 \\ 1 \\ 0 \\ 0 \end{pmatrix}, \begin{pmatrix} 0 \\ 0 \\ 1 \\ 0 \end{pmatrix}, \begin{pmatrix} 0 \\ 0 \\ 0 \\ 1 \end{pmatrix} \tag{5.3.6}$$

are obtained. The energy eigenvalues are mc^2 for two solutions and $-mc^2$ for the other two solutions. Two solutions for each energy eigenvalue are interpreted as two spin states. As shown in the following, the Zeeman energy shifts in opposite directions are derived for both spin states when a magnetic field is applied. The meaning of negative rest energy is discussed in chapter 5.5.

Here, we consider the kinetic energy K with $E \approx mc^2$.

$$E\vec{u} = mc^2\vec{u} + c[\sigma_x\widetilde{p_x} + \sigma_y\widetilde{p_y} + \sigma_z\widetilde{p_z}]\vec{w}$$
$$E\vec{w} = -mc^2\vec{w} + c[\sigma_x\widetilde{p_x} + \sigma_y\widetilde{p_y} + \sigma_z\widetilde{p_z}]\vec{u} \tag{5.3.7}$$

For the limit $K = E - mc^2 \ll mc^2$, eq. (5.3.7) is rewritten as:

$$K\vec{u} = c[\sigma_x\widetilde{p_x} + \sigma_y\widetilde{p_y} + \sigma_z\widetilde{p_z}]\vec{w}$$

using the approximation $(E + mc^2)\vec{w} = 2mc^2\vec{w}$

$$2mc^2\vec{w} = c[\sigma_x\widetilde{p_x} + \sigma_y\widetilde{p_y} + \sigma_z\widetilde{p_z}]\vec{u} \rightarrow \vec{w} = \frac{[\sigma_x\widetilde{p_x}+\sigma_y\widetilde{p_y}+\sigma_z\widetilde{p_z}]}{2mc}\vec{u}$$

based on eq. (5.1.4)

$$K\vec{u} = \frac{\widetilde{p_x}^2+\widetilde{p_y}^2+\widetilde{p_z}^2}{2m}\vec{u} \tag{5.3.8}$$

Equation (5.3.8) corresponds to the Schroedinger equation. The negative value of the kinetic energy is derived from

$$K\vec{w} = -\frac{\widetilde{p_x}^2+\widetilde{p_y}^2+\widetilde{p_z}^2}{2m}\vec{w} \tag{5.3.9}$$

The Schroedinger equation is a second-order differential equation in terms of x, y, and z, whereas the Dirac equation is a first-order differential equation of a vector. The Dirac equation transforms a higher-order differential equation into a multi-dimensional first-order differential equation, as shown in chapter 1. The energy eigenvalue is the same (called degenerated) for both spin states in the absence of an electromagnetic field.

For an electromagnetic field, we use the exchange $E \rightarrow E - q_e\Phi$ and $\vec{p} \rightarrow \vec{p} + q_e\vec{A}$ as shown in chapter 2.4. We obtain $E' = E - mc^2$ using the classical approximation $(E + mc^2 \approx 2mc^2)$ for $\vec{A} = \frac{1}{2}(-yB, xB, 0)$ and $\vec{B} = (0,0,B)$. The product between $\sigma_{x,y}(p_{x,y} + q_eA_{x,y})$ is given by:

$$\left[\sigma_x\left(\frac{h}{2\pi i}\frac{\partial}{\partial x}-\frac{q_e yB}{2}\right)\right]^2$$

$$=\begin{pmatrix}-\left(\frac{h}{2\pi}\right)^2\frac{\partial^2}{\partial x^2}-q_c By\left(\frac{h}{2\pi i}\right)\frac{\partial}{\partial x}+\left(\frac{q_e yB}{2}\right)^2 & 0 \\ 0 & -\left(\frac{h}{2\pi}\right)^2\frac{\partial^2}{\partial x^2}-q_c By\left(\frac{h}{2\pi i}\right)\frac{\partial}{\partial x}+\left(\frac{q_e yB}{2}\right)^2\end{pmatrix}$$

$$\left[\sigma_y\left(\frac{h}{2\pi i}\frac{\partial}{\partial y}+\frac{q_e xB}{2}\right)\right]^2$$

$$=\begin{pmatrix}-\left(\frac{h}{2\pi}\right)^2\frac{\partial^2}{\partial y^2}+q_c Bx\left(\frac{h}{2\pi i}\right)\frac{\partial}{\partial y}+\left(\frac{q_e xB}{2}\right)^2 & 0 \\ 0 & -\left(\frac{h}{2\pi}\right)^2\frac{\partial^2}{\partial y^2}+q_c Bx\left(\frac{h}{2\pi i}\right)\frac{\partial}{\partial y}+\left(\frac{q_e xB}{2}\right)^2\end{pmatrix}$$

$$\sigma_x\left(\frac{h}{2\pi i}\frac{\partial}{\partial x}-\frac{q_e yB}{2}\right)\sigma_y\left(\frac{h}{2\pi i}\frac{\partial}{\partial y}+\frac{q_e xB}{2}\right)+\sigma_y\left(\frac{h}{2\pi i}\frac{\partial}{\partial y}+\frac{q_e xB}{2}\right)\sigma_x\left(\frac{h}{2\pi i}\frac{\partial}{\partial x}-\frac{q_e yB}{2}\right)=\frac{q_e h}{2\pi}B\begin{pmatrix}1 & 0 \\ 0 & -1\end{pmatrix} \quad \textbf{(5.3.10)}$$

Taking

$$L_z=\left[x\left(\frac{h}{2\pi i}\frac{\partial}{\partial y}\right)-y\left(\frac{h}{2\pi i}\frac{\partial}{\partial x}\right)\right] \quad S_z=\frac{h}{4\pi}\sigma_z \quad \textbf{(5.3.11)}$$

$$E'\vec{u}=\left[\frac{\widetilde{p_x}^2+\widetilde{p_y}^2+\widetilde{p_z}^2}{2m}+\frac{q_e}{2m}L_zB+q_c\Phi\right]I\vec{u}+2\frac{q_e}{2m}BS_z\vec{u} \quad \textbf{(5.3.12)}$$

is obtained. For an applied magnetic field, there are Zeeman energy shifts for both spin states in the positive and negative directions with the same absolute value. For an electron, the Zeeman energy shift is described by

$$L_z=\frac{h}{2\pi}M_L, \; S_z=\frac{h}{2\pi}M_S$$

$$E_Z=\mu_B[M_L+g_SM_S]B \qquad \mu_B=\frac{eh}{4\pi m_e} \quad \textbf{(5.3.13)}$$

The g-factor of the electron spin g_S is obtained as exact 2 from eqs. (5.3.10) and (5.3.12), but it was experimentally determined to be 2.0023 [7]. This discrepancy is the "anomalous magnetic moment", which is derived based on the quantum electrodynamics [8] treatment of the energy fluctuations in a vacuum.

There is also a spin associated with other particles. The Zeeman energy shift induced by the nuclear spin is much smaller than that of the electron because of the larger mass. However, the magnetic dipole moment is not proportional to $\frac{q_c}{m}$. For example, there is a Zeeman energy shift associated with the neutrons.

The existence of the Zeeman energy shift induced by the electron spin is not a relativistic effect, but it was obtained from the Dirac equation, which requires the Hamiltonian to be treated as matrices.

5.4. RELATIVISTIC CORRECTION USING THE DIRAC EQUATION

Chapter 5.3 considered the classical approximation, taking $2mc^2\vec{w} = c\big[\sigma_x\widetilde{p_x} + \sigma_y\widetilde{p_y} + \sigma_z\widetilde{p_z}\big]\vec{u}$. Assuming $\vec{A} = 0$ and $\Phi \neq 0$, relativistic effects are derived by performing a more detailed analysis, as follows:

$$(2mc^2 + E')\vec{w} = c\big[\sigma_x\widetilde{p_x} + \sigma_y\widetilde{p_y} + \sigma_z\widetilde{p_z}\big]\vec{u}$$

$$\vec{w} = \frac{c\big[\sigma_x\widetilde{p_x}+\sigma_y\widetilde{p_y}+\sigma_z\widetilde{p_z}\big]\vec{u}}{2mc^2+E'} = \frac{c\big[\sigma_x\widetilde{p_x}+\sigma_y\widetilde{p_y}+\sigma_z\widetilde{p_z}\big]\vec{u}}{2mc^2} - \frac{E'c\big[\sigma_x\widetilde{p_x}+\sigma_y\widetilde{p_y}+\sigma_z\widetilde{p_z}\big]\vec{u}}{(2mc^2)^2}$$

$$E'\vec{u} = E'_{cl}\vec{u} + E'_{rel}\vec{u}$$

$$E'_{cl} = \frac{\widetilde{p_x}^2 + \widetilde{p_y}^2 + \widetilde{p_z}^2}{2m} + q_e\Phi$$

$$E'_{rel} = -c\big[\sigma_x\widetilde{p_x} + \sigma_y\widetilde{p_y} + \sigma_z\widetilde{p_z}\big]\frac{E'c\big[\sigma_x\widetilde{p_x}+\sigma_y\widetilde{p_y}+\sigma_z\widetilde{p_z}\big]}{(2mc^2)^2} \tag{5.4.1}$$

To calculate the relativistic correction, E'_{rel} was calculated by taking $E' = E'_{cl}$ for simplicity. Then E'_{rel} is separated into:

$$E'_{rel} = K_{rel} + P_{rel} \tag{5.4.2}$$

K_{rel} is the relativistic correction of the kinetic energy, which is given by

$$K_{rel} = -c\big[\sigma_x\widetilde{p_x} + \sigma_y\widetilde{p_y} + \sigma_z\widetilde{p_z}\big]\frac{\widetilde{p_x}^2 + \widetilde{p_y}^2 + \widetilde{p_z}^2}{2m}\frac{c\big[\sigma_x\widetilde{p_x}+\sigma_y\widetilde{p_y}+\sigma_z\widetilde{p_z}\big]}{(2mc^2)^2} = -\frac{\widetilde{p_x}^4 + \widetilde{p_y}^4 + \widetilde{p_z}^4}{8m^3c^2} \tag{5.4.3}$$

This term is derived as the third term of the energy with $\Phi = 0$:

$$E = mc^2\sqrt{1 + \left(\frac{|\vec{p}|}{mc}\right)^2} = mc^2 + \frac{|\vec{p}|^2}{2m} - \frac{|\vec{p}|^4}{8m^3c^2} \tag{5.4.4}$$

The relativistic effect induced by Φ is given by:

$$P_{rel} = -\left[\sigma_x \widetilde{p_x} + \sigma_y \widetilde{p_y} + \sigma_z \widetilde{p_z}\right] q_e \Phi \frac{\left[\sigma_x \widetilde{p_x} + \sigma_y \widetilde{p_y} + \sigma_z \widetilde{p_z}\right]}{4m^2 c^2} \tag{5.4.5}$$

Assuming that Φ is spherically symmetric,

$$\sigma_x \widetilde{p_x}\Phi = \begin{pmatrix} 0 & \frac{h}{2\pi i}\frac{\partial \Phi}{\partial x} \\ \frac{h}{2\pi i}\frac{\partial \Phi}{\partial x} & 0 \end{pmatrix} = \begin{pmatrix} 0 & \frac{h}{2\pi i}\frac{\partial \Phi}{\partial r}\frac{x}{r} \\ \frac{h}{2\pi i}\frac{\partial \Phi}{\partial r}\frac{x}{r} & 0 \end{pmatrix}$$

$$\sigma_y \widetilde{p_y}\Phi = \begin{pmatrix} 0 & -\frac{h}{2\pi}\frac{\partial \Phi}{\partial r}\frac{y}{r} \\ \frac{h}{2\pi}\frac{\partial \Phi}{\partial r}\frac{y}{r} & 0 \end{pmatrix} \quad \sigma_z \widetilde{p_z}\Phi = \begin{pmatrix} \frac{h}{2\pi i}\frac{\partial \Phi}{\partial r}\frac{z}{r} & 0 \\ 0 & -\frac{h}{2\pi i}\frac{\partial \Phi}{\partial r}\frac{z}{r} \end{pmatrix}$$

$$\sigma_x \widetilde{p_x}\Phi\left[\sigma_x \widetilde{p_x} + \sigma_y \widetilde{p_y} + \sigma_z \widetilde{p_z}\right] = -\left(\frac{h}{2\pi}\right)^2 \frac{\partial \Phi}{r\partial r}\begin{pmatrix} x\left(\frac{\partial}{\partial x}+i\frac{\partial}{\partial y}\right) & -x\frac{\partial}{\partial z} \\ x\frac{\partial}{\partial z} & x\left(\frac{\partial}{\partial x}-i\frac{\partial}{\partial y}\right) \end{pmatrix}$$

$$\sigma_y \widetilde{p_y}\Phi\left[\sigma_x \widetilde{p_x} + \sigma_y \widetilde{p_y} + \sigma_z \widetilde{p_z}\right] = -\left(\frac{h}{2\pi}\right)^2 \frac{\partial \Phi}{r\partial r}\begin{pmatrix} -iy\left(\frac{\partial}{\partial x}+i\frac{\partial}{\partial y}\right) & -iy\frac{\partial}{\partial z} \\ iy\frac{\partial}{\partial z} & iy\left(\frac{\partial}{\partial x}-i\frac{\partial}{\partial y}\right) \end{pmatrix}$$

$$\sigma_z \widetilde{p_z}\Phi\left[\sigma_x \widetilde{p_x} + \sigma_y \widetilde{p_y} + \sigma_z \widetilde{p_z}\right] = -\left(\frac{h}{2\pi}\right)^2 \frac{\partial \Phi}{r\partial r}\begin{pmatrix} z\frac{\partial}{\partial z} & z\left(\frac{\partial}{\partial x}-i\frac{\partial}{\partial y}\right) \\ -z\left(\frac{\partial}{\partial x}+i\frac{\partial}{\partial y}\right) & z\frac{\partial}{\partial z} \end{pmatrix}$$

$$P_{rel} = P_d + P_{fs}$$

$$P_d = \frac{q_e}{4m^2 c^2}\left(\frac{h}{2\pi}\right)^2 \frac{\partial \Phi}{r\partial r}\begin{pmatrix} x\frac{\partial}{\partial x}+y\frac{\partial}{\partial y}+z\frac{\partial}{\partial z} & 0 \\ 0 & x\frac{\partial}{\partial x}+y\frac{\partial}{\partial y}+z\frac{\partial}{\partial z} \end{pmatrix} = \frac{q_e}{4m^2 c^2}\left(\frac{h}{2\pi}\right)^2 \frac{\partial^2 \Phi}{\partial r^2} I$$

$$P_{fs} = \frac{q_e}{4m^2 c^2}\frac{h}{2\pi i}\frac{\partial \Phi}{r\,\partial r}\left[L_x \sigma_x + L_y \sigma_y + L_z \sigma_z\right] = \frac{q_e}{2m^2 c^2}\frac{\partial \Phi}{r\,\partial r}\vec{L}\cdot\vec{S} \tag{5.4.6}$$

Here, P_d is the Darwin's term, and P_{fs} is the fine structure term, given as the coupling between the orbital angular momentum and the spin. Using the sum angular momentum $\vec{J} = \vec{L} + \vec{S}$,

$$\vec{L}\cdot\vec{S} = \frac{J(J+1)-L(L+1)-S(S+1)}{2} \text{ (from the cosine theorem)} \tag{5.4.7}$$

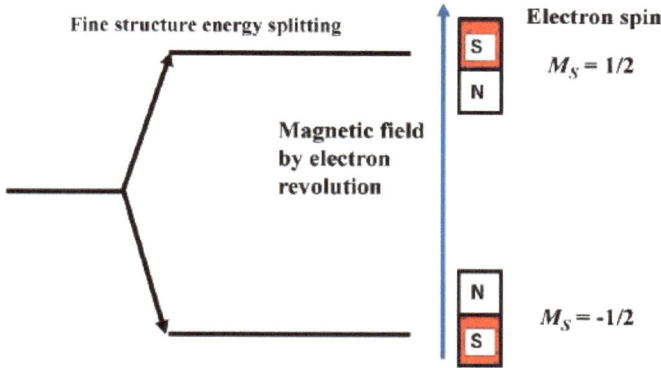

Fig. (5.1). Fine structure energy splitting by the magnetic field induced by the electron revolution and the electron spin. (This figure is used also in "Cold Atoms and Molecules" by M. Kajita).

For an electron with $S = \frac{1}{2}$, the $J = L + \frac{1}{2}$ and $J = L - \frac{1}{2}$ states denote the states in which the orbital angular momentum and spin are parallel and antiparallel, respectively. For an electron in a hydrogen-like ion (one electron and the nuclear with a charge of $+Ze$),

$$P_{fs} = \frac{1}{16\pi\varepsilon_0 m_e^2 c^2} \frac{Ze^2}{r^3} [J(J+1) - L(L+1) - S(S+1)] \qquad (5.4.8)$$

which gives the energy gap between different J states in the same (n, L) state. This energy splitting is induced by the interaction between the electron spin and the magnetic field induced by the electron's revolution motion, which is parallel or antiparallel (see Fig. **5.1**). Including the fine structure, the electron energy state of an atom is described by:

$$L = 0 \quad n^{2S+1}S_J,$$
$$L = 1 \quad n^{2S+1}P_J,$$
$$L = 2 \quad n^{2S+1}D_J$$

The energy eigenvalue of the electron in a hydrogen atom or hydrogen like ions (one electron and the nucleus with a charge of $+Ze$) is obtained strictly from the Dirac equation as

$$E_{n,J} = \frac{m_e c^2}{\sqrt{1+\left(\dfrac{Z\alpha}{n-J-\frac{1}{2}+\sqrt{\left(J+\frac{1}{2}\right)^2-(Z\alpha)^2}}\right)^2}}}$$

$$\alpha = \frac{e^2}{2\varepsilon_0 hc} \quad \text{(fine structure constant)} \tag{5.4.9}$$

For the value of $\alpha = 0.00116$, the following approximation is valid for a small Z.

$$\sqrt{\left(J+\frac{1}{2}\right)^2-(Z\alpha)^2} = \left(J+\frac{1}{2}\right) - \frac{(Z\alpha)^2}{2\left(J+\frac{1}{2}\right)}$$

$$E_{n,J} = \frac{m_e c^2}{\sqrt{1+\left(\dfrac{Z\alpha}{n-\frac{(Z\alpha)^2}{2\left(J+\frac{1}{2}\right)}}\right)^2}} = m_e c^2 - \frac{m_e c^2 Z^2 \alpha^2}{2\left(n-\frac{(Z\alpha)^2}{2\left(J+\frac{1}{2}\right)}\right)^2}$$

$$\text{taking } \frac{1}{\left(n-\frac{(Z\alpha)^2}{2\left(J+\frac{1}{2}\right)}\right)^2} = \frac{1}{n^2} + \frac{(Z\alpha)^2}{n^3\left(J+\frac{1}{2}\right)}$$

$$= m_e c^2 - \frac{m_e Z^2 e^4}{8\varepsilon_0^2 h^2 n^2} - \frac{m_e Z^4 e^4}{8\varepsilon_0^2 h^2 n^3}\frac{\alpha^2}{\left(J+\frac{1}{2}\right)} \tag{5.4.10}$$

The first term denotes the rest energy of an electron. The second term represents the energy eigenvalue obtained using the Schroedinger equation, and the third term denotes the relativistic effect depending on J. Note that the relative motion between the electron and nucleus, and the motion of the center of mass are not separable based on the relativistic treatment. Therefore, it is often treated by changing m_e to μ_e (reduced mass) in the second and third terms of eq. (5.4.10). The dependence of the energy eigenvalues on α is significant for highly charged ions. By measuring the ratio between the transition frequencies of atoms or ions with small Z and large Z, we can determine if there is a temporal variation in α [9].

There is an energy gap between the $2^2P_{3/2}$ and $2^2P_{1/2}$ states with a transition frequency of 10.9 GHz for hydrogen atoms. The Nucleus also has a spin, and the interaction between the electron spin and the electron orbital angular momentum induces another energy splitting, called hyperfine splitting. The transition frequency of the hyperfine splitting of hydrogen atoms in the $^1S_{1/2}$ state is 1.4 GHz.

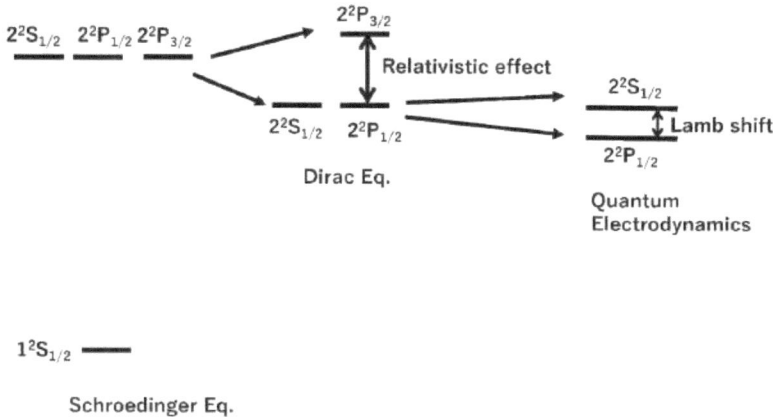

Fig. (5.2). The energy structure of H atom according to the Schroedinger and the Dirac equations and quantum electrodynamics. (This figure is used also in "Cold Atoms and Molecules" by M. Kajita).

Fig. (**5.2**) shows the electron energy state given by the Schroedinger equation and Dirac equations. Although the Dirac equation can be solved to obtain the energy eigenvalues, there are still discrepancies with the experimental results. For example, the $2^2S_{1/2}$ and $2^2P_{1/2}$ states should have the same energy according to the Dirac equation. However, a slight energy gap between both states (called the Lamb shift) was observed, as shown in Fig. (**5.2**) [10]. This energy gap can be explained by the interaction between the vacuum energy fluctuations and the electrons in different orbits ($2^2S_{1/2}$: spherical symmetric, $2^2P_{1/2}$: polarized). The discovery of this slight energy gap played a significant role in developing the field of "quantum electrodynamics" considering vacuum energy fluctuations [8]. The energy in the $1^2S_{1/2}$ state is also shifted from the solution of the Dirac equation by the quantum electromagnetic effect. There is currently no discrepancy between the measurement and calculation, which is larger than the uncertainties.

The remaining source of discrepancies between the experiment and calculation arises from the finite size of the proton. For measurement using hydrogen atoms, it was estimated to be 0.8775 ± 0.005 fm [11]. However, the result obtained for a muonic hydrogen atom (proton + muon) was 0.842 ± 0.001 fm [12]. The discrepancy in the results is five times larger than the measurement uncertainty and, thus, is a mystery of modern physics called the "proton radius puzzle".

5.5. MEANING OF NEGATIVE REST ENERGY

As shown in chapter 5.3, negative energy eigenvalues are obtained for the Dirac equation. To solve this mystery, Dirac proposed the idea of the electron sea, a vacuum filled with electrons with negative energy [13]. Given that the electron is a Fermion (only one particle can exist in a single quantum state, as shown in (chapter 6.6), electrons with positive energy cannot decay to the negative energy state when all the quantum states in the vacuum are filled with electrons with negative energy.

For energy of $2m_ec^2$ owing to a γ- ray, electrons with negative energy are excited to the positive energy state, and we recognize them as the appearance of conventional electrons, as shown in Fig. (**5.3**). The absence of electrons with negative energy is observed as a positively charged particle with the same mass as the electron. This idea was confirmed by the discovery of a positron [14]. The pair production and extinction of electrons and positrons were also observed. Antiprotons and antineutrons have also been discovered. There are also antiparticles whose characteristics are almost the same except for the inverse electric charge for all particles (the equality of characteristics should not be perfect as shown in chapter 7).

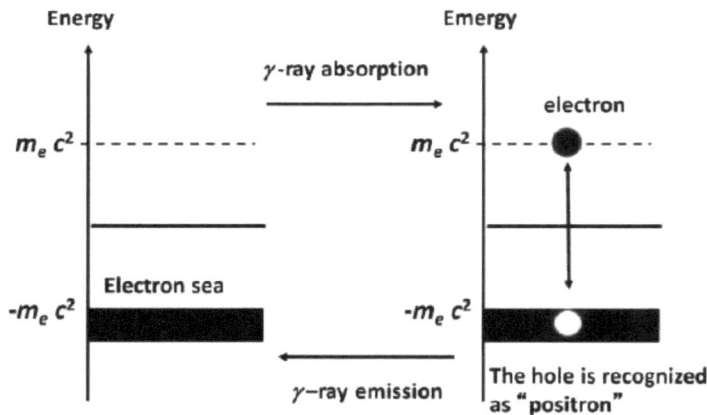

Fig. (5.3). The idea of "electron sea model" given by Dirac. The existence of positron was discovered afterwards. (This figure is used also in "Measuring Time: Frequency measurement and related developments in physics" by M. Kajita).

If we apply the idea of the electron sea for Boson (no limit on the number of particles in the same quantum state (chapter 6.6), all particles can be in the negative energy state. For the existence of particles with positive rest energy, the existence of a negative rest energy state is not possible for Boson. However, antiparticles have also been discovered with Boson. Therefore, the idea of an electron sea is not currently used. A new interpretation was provided by Feynman [15].

5.6. UTILITY OF THE EQUATIONS FOR WAVE FUNCTIONS WITH DIFFERENT SPIN

The Dirac equation using a 4×4 matrix is applicable for the particles with $S = \frac{1}{2}$. For the particles with $S = 0$, the wave function is scalar, and the Klein-Gordon equation is applicable [4]. For the particles with other spin values, the following equations are applicable.

The Rarita-Schwiner equation   massive particles with $S = \frac{3}{2}$ [16]
The Bargmann-Wigner equation   free particles with arbitral spin [17]

The Bargmann-Wigner equation is a set of linear partial differential equations with $2S$ components. The calculation of each component was based on the Dirac equation ($2S = 1$). However, it is difficult to use the Bargmann-Wigner equation for particles in an electromagnetic field.

EXERCISE

Show that the square of the Hamiltonian given by the Dirac equation corresponds to the Klein-Gordon equation. Moreover, show that the Dirac equation converges to the Schrödinger equation in the limit of $mc^2 \gg c|\vec{p}|$, except for the inclusion of the Zeeman energy shift by the electron spin.

REFERENCES

[1]   W. Gerlach, and O. Stern, "Der experimentalle Nachweise der Richtungsquantelung im MAgnetfeld", *Eur. Phys. J. A,* vol. 9, no. 1, pp. 349-352, 1922.
[http://dx.doi.org/10.1007/BF01326983]

[2]   R.L. Liboff, *Introductory Quantum Mechanics.* Addison-Wesley, 2002.Pauli matrices - Wikipedia

[3]   O. Klein, "Quanten theorie und fuenfdimensionale Relativitaettheorie", *Eur. Phys. J. A,* vol. 37, no. 12, pp. 895-906, 1926.
[http://dx.doi.org/10.1007/BF01397481]

[4]   W. Pauli, and V. Weisskopf, "On Quantization of the Relativistic Wave Equation", *Helv. Phys. Acta,* vol. 7, pp. 709-731, 1934.

[5]   P.A.M. Dirac, "Bakerian Lecture. The Physical Interpretation of Quantum Mechanics", *Proc.- Royal Soc., Math. Phys. Eng. Sci.,* vol. 180, no. 980, pp. 1-39, 1942.

[6]   P.A.M. Dirac, *Principles of Quantum Mechanics. International Series of Monographs on Physics.* 4th ed. Oxford University Press, 1958, p. 255.ISBN978-0-19-852011-5

[7]   R. Kusch, and H.M. Foley, "The Magnetic Moment of the Electron", *Phys. Rev.,* vol. 74, no. 3, pp. 250-263, 1948.
        [http://dx.doi.org/10.1103/PhysRev.74.250]

[8]   J. Schwinger, "On Quantum-Electrodynamics and the Magnetic Moment of the Electron", *Phys. Rev.,* vol. 73, no. 4, pp. 416-417, 1948.
        [http://dx.doi.org/10.1103/PhysRev.73.416]

[9]   M. Kajita, *Measurement, Uncertainties and Lasers.* IOP Expanding Physics, 2019, pp. 12-13.
        [http://dx.doi.org/10.1088/2053-2563/ab0373]

[10]   W.E. Lamb, and R.C. Retherford, "Fine Structure of the Hydrogen Atom by Microwave Method", *Phys. Rev.,* vol. 72, no. 3, pp. 241-243, 1947.
        [http://dx.doi.org/10.1103/PhysRev.72.241]

[11]   I. Sick, and D. Trautmann, "Proton root-mean-square radii and electron scattering", *Phys. Rev. C Nucl. Phys.,* vol. 89, no. 1, 2014.012201 [R].
        [http://dx.doi.org/10.1103/PhysRevC.89.012201]

[12]   R. Pohl, A. Antognini, F. Nez, *et al.,* "The size of the proton", *Nature,* vol. 466, no. 7303, pp. 213-216, 2010. [http://dx.doi.org/10.1038/nature09250] [PMID: 20613837]

[13]   P.A.M. Dirac, "A theory of electrons and protons", *Proc. R. Soc. Lond., A Contain. Pap. Math. Phys. Character,* vol. 126, no. 801, pp. 360-365, 1930.
        [http://dx.doi.org/10.1098/rspa.1930.0013]

[14]   Close F. Antimatter. Oxford U. Press 2009; 50, ISBN 978-0-19-955016-6.

[15]   R.P. Feynman, "Space-Time Approach to Non-Relativistic Quantum Mechanics", *Rev. Mod. Phys.,* vol. 20, no. 2, pp. 367-402, 1948.
        [http://dx.doi.org/10.1103/RevModPhys.20.367]

[16]   W. Rarita, and J. Schwiner, "On a theory of particles with half-integral spin", *Phys. Rev.,* vol. 60, no. 1, p. 61, 1941.
        [http://dx.doi.org/10.1103/PhysRev.60.61]

[17]   V. Bargmann, and E.P. Wigner, "Group theoretical discussion of relativistic wave equations", *Proc. Natl. Acad. Sci. USA,* vol. 34, no. 5, pp. 211-223, 1948.
        [http://dx.doi.org/10.1073/pnas.34.5.211] [PMID: 16578292]

CHAPTER 6

Fundamentals of Statistical Mechanics Using the Boltzmann Distribution

Abstract: The objective of this chapter is an examination of the energy distribution of matter for the highest probability, considering that only phenomena with the highest possibility are possible for a large number of masses (atoms or molecules in a macroscopic object). The thermodynamic temperature T is the parameter for the broadening of the energy distribution and the population in a state with an energy of E is proportional to $\Omega \exp(-E/(k_B T))$, where Ω is the number of states and k_B is the Boltzmann constant. The average energy and specific heat are discussed using the Boltzmann distribution. The relationship between the gas pressure, volume, and temperature (ideal gas law) is obtained from the average of the one-dimensional kinetic energy. The work efficiency of the Carnot engine, using gas pressure, is also discussed.

Keywords: Adiabatic expansion, Bose-Einstein condensation, Carnot cycle, Entropy, Fermi degeneracy, Free energy, Ideal gas law, Specific heat, Temperature, Thermal energy, Thermal equilibrium, Vapor pressure.

6.1. THERMAL ENERGY

What is thermal energy? From a microscopic perspective, the constituent atoms or molecules of an object have kinetic and potential energies. However, at the macroscopic scale, the kinetic energy of the object is given by that of the motion of the center of mass (see chapter 2.2). The relative motion between the constituent atoms or molecules provides an additional energy term called thermal energy. For example, we consider the kinetic energy of gaseous molecules with a mass of m. The total energy is given by

$$E_{tot} = \frac{m}{2}\sum_{i=1}^{N_g} v_i^2 \qquad N_g : \text{number of molecules} \tag{6.1.1}$$

Considering the average velocity

$$v_{ave} = \frac{\sum_{i=1}^{N_g} v_i}{N} \quad v_i = v_{ave} + \delta v_i$$

$$E_{tot} = E_K + E_T \quad (\text{note } \sum_{i=1}^{N_g} v_{ave}(\delta v_i) = 0)$$

$$E_K = \frac{N_g m}{2} v_{ave}^2 \quad E_T = \frac{m}{2} \sum_{i=1}^{N_g} (\delta v_i)^2 \tag{6.1.2}$$

is obtained. E_K is the macroscopic kinetic energy and E_T is the thermal energy.

The macroscopic kinetic energy tends to transform into thermal energy. When an object slides on a floor, it is decelerated by friction, and the kinetic energy is transformed into thermal energy, which expands into the entire floor. Why does inverse energy flow (the matter on the floor start to move to absorb the thermal energy from the floor) not occur? Statistic mechanics dictates that the probability of the expansion of thermal energy is much higher than that of focusing on one location. When two bodies with different temperatures contact each other, the energy (heat) flows from the body at a higher temperature to another body at a lower temperature. Finally, the temperatures of both bodies become equal, for which the probability is maximum.

6.2. BOLTZMANN DISTRIBUTION

We consider two areas A and B with energies E_A and E_B, for which $E_{tot} = E_A + E_B$ is constant. Considering the number of states $\Omega_A(E_A)$ and $\Omega_B(E_B)$ at both areas, the probability of achieving this energy distribution is proportional to the total number of states given by:

$$\Omega_{tot} = \Omega_A(E_A)\Omega_B(E_B) \tag{6.2.1}$$

When Ω_{tot} is a maximum, the following relation holds.

$$\frac{d\Omega_{tot}}{dE_A} = \Omega_B(E_B)\frac{d\Omega_A(E_A)}{dE_A} + \Omega_A(E_A)\frac{d\Omega_B(E_B)}{dE_A} = \Omega_B(E_B)\frac{d\Omega_A(E_A)}{dE_A} - \Omega_A(E_A)\frac{d\Omega_B(E_B)}{dE_B} = 0$$

$$\frac{1}{\Omega_A(E_A)}\frac{d\Omega_A(E_A)}{dE_A} = \frac{1}{\Omega_B(E_B)}\frac{d\Omega_B(E_B)}{dE_B} = \beta \tag{6.2.2}$$

Equation (6.2.2) shows that the probability is a maximum when the parameter β, defined in A and B independently, is equal. Note also that the dimension β is the inverse of that of energy. We know empirically that the temperatures of the two contacting bodies become equal. In statistical mechanics, the thermodynamic temperature T is defined as:

$$\frac{1}{k_B T} = \beta = \frac{1}{\Omega(E)}\frac{d\Omega(E)}{dE} \tag{6.2.3}$$

Here, k_B is the Boltzmann constant, defined as $1.38064852 \times 10^{-23}$ J/K in 2019 [1]. The statistical mechanical entropy is defined as:

$$S = k_B \ln(\Omega) \tag{6.2.4}$$

and eqs. (6.2.1-2) are rewritten as:

$$S_{tot} = S_A + S_B \tag{6.2.5}$$

$$\frac{1}{T} = \frac{dS}{dE} \tag{6.2.6}$$

We can estimate Ω_B using β and E_A as follows:

$$\frac{1}{\Omega_B(E_B)}\frac{d\Omega_B(E_B)}{dE_B} = -\frac{1}{\Omega_B(E_B)}\frac{d\Omega_B(E_B)}{dE_A} = \beta$$
$$\int \frac{1}{\Omega_B}\frac{d\Omega_B}{dE_A} dE_A = -\beta \int dE_A$$
$$\ln(\Omega_B) = -\beta E_A + const$$
$$\Omega_B \propto \exp(-\beta E_A) = \exp\left(-\frac{E_A}{k_B T}\right) \tag{6.2.7}$$

Then eq. (6.2.1) is rewritten as:

$$\Omega_{tot} \propto \Omega_A(E_A)\exp\left(-\frac{E_A}{k_B T}\right) \tag{6.2.8}$$

This energy distribution is called the "Boltzmann distribution". For example, the gravitational potential distribution is proportional to $\exp\left(-\frac{mgh_g}{k_B T}\right)$ (g: gravitational acceleration, h_g: height from the ground) and the mass can be distributed within the region:

$$0 \leq h_g < \frac{k_B T}{mg} \tag{6.2.9}$$

For an object with a mass of 1 g (10^{-3} kg), the possible floating height is 4×10^{-19} m, which is much less than the size of a proton, and floating is impossible. However, an oxygen (O_2) molecule (5.3×10^{-26} kg) is distributed between $h_g = 0$ and 7900 m.

The distribution at h_g = 7900 m is 0.36 times lower than that at the surface. This estimation is reasonable compared to the altitude-oxygen chart [2].

The probability that area A requires energy E_A is given by:

$$\rho_P(E_A)dE_A = \frac{\Omega_A(E_A)\exp(-\beta E_A)dE_A}{Z}$$

$$Z = \int \Omega_A(E_A)\exp(-\beta E_A)\,dE_A \tag{6.2.10}$$

Here, Z is the normalizing parameter required to satisfy

$$\int \rho_P(E_A)dE_A = 1 \tag{6.2.11}$$

6.3. AVERAGE OF ENERGY

The average of the energy is given by:

$$\langle E\rangle_{ave} = \int E\rho_P(E)dE = \frac{\int E\Omega\exp(-\beta E)dE}{Z} \tag{6.3.1}$$

Using

$$\frac{d}{d\beta}\exp(-\beta E) = -E\exp(-\beta E) \tag{6.3.2}$$

eq. (6.3.1) can be rewritten as:

$$\langle E\rangle_{ave} = -\frac{1}{Z}\frac{dZ}{d\beta} = -\frac{d}{d\beta}\ln(Z) \tag{6.3.3}$$

The average one-dimensional kinetic energy ($K = mv^2/2$, where m is the mass and v is the velocity) is given by:

$$Z = \int \exp\left(-\frac{m\beta}{2}v^2\right)dv.$$

$$\text{using } p_\alpha = \sqrt{\beta m/2}\,v$$

$$Z = \frac{dv}{dp_\alpha} P_1 = \sqrt{\frac{2}{m\beta}}\, P_1 \qquad P_1 = \int \exp(-p_\alpha^2) dp_\alpha$$

$$\ln Z = \ln(P_1) + \frac{1}{2}\ln\left(\frac{2}{m}\right) - \frac{1}{2}\ln(\beta)$$

$$\langle K_E \rangle_{ave} = \frac{m}{2}\langle v^2 \rangle = -\frac{d}{d\beta}\ln(Z) = \frac{1}{2\beta} = \frac{k_B T}{2} \tag{6.3.4}$$

Height $k_B T/2mg$

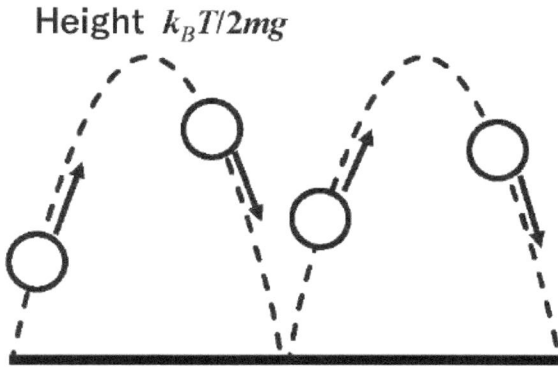

Fig. (6.1). Motion of gaseous atoms and molecules, bounding at surface. The launching height is given by $k_B T/2mg$, where T is the thermodynamic temperature, k_B is the Boltzmann constant, m is the mass, and g is the gravitational acceleration, respectively. (This figure is used also in "Cold Atoms and Molecules" by M. Kajita).

The average of the velocity one direction is given by:

$$|v| = \sqrt{\frac{k_B T}{m}}. \tag{6.3.5}$$

The launching height with this initial velocity given by eq.(6.3.5) is:

$$h_{launch} = \frac{v^2}{2g} = \frac{k_B T}{2mg} \tag{6.3.6}$$

which is lower than the area given by eq. (6.2.6) with a factor of $\frac{1}{2}$. Gaseous atoms and molecules repeat bounding at the surface, as shown in Fig. (**6.1**).

Equation (6.3.4) shows that the average kinetic energy is independent of the mass. However, we can consider that matter can consist of several mass components. Without binding, each mass has kinetic energy given by eq. (6.3.4) and can move with higher velocity. Why does velocity become slower owing to binding? The average velocity of the center of mass of two particles with masses m_1 and m_2 is given by (see chapter 2.2):

$$v_c = \frac{m_1 v_1 + m_2 v_2}{m_1 + m_2}$$

$$\langle v_c^2 \rangle_{ave} = \frac{m_1^2 \langle v_1^2 \rangle_{ave} + 2 m_1 m_2 \langle v_1 v_2 \rangle_{ave} + m_2^2 \langle v_2^2 \rangle_{ave}}{(m_1 + m_2)^2}$$

taking $\langle v_{1,2}^2 \rangle_{ave} = \frac{k_B T}{m_{1,2}}$ $\langle v_1 v_2 \rangle_{ave} = 0$

$$= \frac{k_B T}{m_1 + m_2} \tag{6.3.7}$$

and eq. (6.3.6) is also valid after the binding of the particles. The energy of the relative motion should be considered taking the binding potential energy into account.

Using eq. (1.3.27), the average three-dimensional kinetic energy is obtained as follows:

$$Z = \iiint \exp\left[-\frac{m}{2}\left(v_x^2 + v_y^2 + v_z^2\right)\right] dv_x dv_y dv_z = 4\pi \int v^2 \exp\left(-\frac{m\beta}{2} v^2\right) dv$$

using $p_\alpha = \sqrt{\beta m / 2}\, v$

$$Z = \left(\frac{2}{m\beta}\right)^{3/2} P_3 \qquad P_3 = 4\pi \int p_\alpha^2 \exp(-p_\alpha^2) dp$$

$$\ln Z = \ln(P_3) + \frac{3}{2}\ln\left(\frac{2}{m}\right) - \frac{3}{2}\ln(\beta)$$

$$\langle K_E \rangle_{ave} = \frac{m}{2}\langle v^2 \rangle = -\frac{d}{d\beta}\ln(Z) = \frac{3}{2\beta} = \frac{3 k_B T}{2} \tag{6.3.8}$$

This result indicates that the three-dimensional kinetic energy is the sum of the kinetic energies in the three directions (in each direction $k_B T/2$). The average potential energy $P_E = Cr^n$ (r: distance) is derived as follows, using the same derivations of eqs. (6.3.4) and (6.3.8):

$$\text{One-dimension } \langle P_E \rangle_{ave} = \frac{k_B T}{n}$$
$$\text{Three-dimension } \langle P_E \rangle_{ave} = \frac{3k_B T}{n} \quad \text{(6.3.9)}$$

From quantum mechanics (see chapter 4), the energy in a bounded state can only be discrete values E_i, and eq. (6.2.10) can be rewritten as:

$$Z = \sum \Omega(E_i) \exp(-\beta E_i) \quad \text{(6.3.10)}$$

For example, when the possible energy is only an integer multiple of unit energy ϵ ($E = n\epsilon$) and $\Omega(n\epsilon) = 1$,

$$Z = \sum \exp(-n\beta\epsilon) = \frac{1}{1 - \exp(-\beta\epsilon)}$$
$$\langle E \rangle_{ave} = -\frac{1}{Z}\frac{dZ}{d\beta} = \frac{\epsilon}{\exp\left(\frac{\epsilon}{k_B T}\right) - 1} \quad \text{(6.3.11)}$$

When $\epsilon \ll k_B T$, $\langle E \rangle_{ave} \approx k_B T$ is obtained using the approximation $\exp\left(\frac{\epsilon}{k_B T}\right) - 1 \approx \frac{\epsilon}{k_B T}$. When $\epsilon \gg k_B T$, $\langle E \rangle_{ave}$ is possible only in the lowest energy state. Fig. (**6.2**) shows the relation between $\frac{k_B T}{\epsilon}$ and $\frac{\langle E \rangle_{ave}}{\epsilon}$.

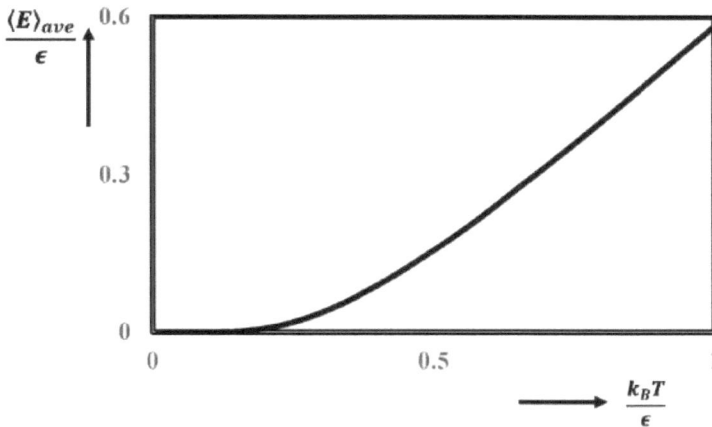

Fig. (6.2). The dependence of the average energy $\langle E \rangle_{ave}$ on $k_B T$ when the possible energy is integer multiple of ϵ. Both values are normalized with ϵ.

6.4. CHANGE OF THE TOTAL ENERGY BY CHANGING THE POTENTIAL PARAMETER

As shown in chapter 2.1, the energy changes with temporal changes in the potential parameters. When a particle is bounded by a potential field with $P_E = Cr^n$, the total energy is given by:

$$E_{tot} = K_E + P_E \tag{6.4.1}$$

The ratio between the average values of K_E and P_E from eqs. (6.3.4-9) is given by:

$$\langle K_E \rangle_{ave} = \frac{n}{n+2} E_{tot} \quad \langle P_E \rangle_{ave} = \frac{2}{n+2} E_{tot} \tag{6.4.2}$$

When C changed to $C + \Delta C$,

$$\frac{\Delta P_E}{P_E} = \frac{\Delta C}{C} \tag{6.4.3}$$

Taking $\Delta P_E = \Delta E_{tot}$ and $P_E = \frac{2}{n+2} E_{tot}$, eq. (6.4.3) is rewritten as:

$$\frac{\Delta E_{tot}}{E_{tot}} = \frac{2}{n+2} \frac{\Delta C}{C}$$

$$E_{tot} \propto C^{\frac{2}{n+2}} \tag{6.4.4}$$

For example, the harmonic oscillation energy ($n = 2$) is proportional to $\sqrt{C}$.

6.5. THE SPECIFIC HEAT OF GASEOUS AND SOLID ATOMS AND MOLECULES

The specific heat at constant volume (heat does not perform any work on the outer system) c_v of matter constructed by bonding the multiple masses is given by:

$$c_v = \frac{d\langle E_{tot} \rangle_{ave}}{dT} = \frac{d(\langle E_c \rangle_{ave} + \langle E_r \rangle_{ave})}{dT}$$

E_c: kinetic energy of the center of mass
E_r: relative motion energy (kinetic energy + potential energy)   (6.5.1)

As shown in eq. (6.3.8), $\langle E_c \rangle_{ave} = \frac{3}{2} k_B T$. The relative motion in the direction parallel and perpendicular to the bonding force is the vibration and rotation, respectively. As shown in chapter 4.4, the vibrational, rotational energies can only assume discrete values, and E_r is constant at the lowest state when $k_B T$ is much smaller than the energy gap.

For the gaseous atoms (He, Ne, Ar, Kr, and Xe), we can consider with $E_{tot} = E_c$ and

$$c_v = \frac{3k_B}{2} \tag{6.5.2}$$

This estimation is in good agreement with experimental results because the energy gap between different electron energy states is much larger than $k_B T$ when $T < 10^5$ K.

For gaseous molecules, the energy gap between different relative interatomic motions (vibrational motion in one direction and rotational motion in two directions) are comparable to $k_B T$ for T = 100-400 K, and c_v has a complicated dependence on T. For the diatomic molecules (see chapter 4.4),

$$\langle E_r \rangle_{ave} = \langle E_{rot} \rangle_{ave} + \langle E_{vib} \rangle_{ave}$$

$$\langle E_{rot} \rangle_{ave} = 2 \frac{\sum (2N+1)N(N+1)B_0 \exp\left(-\frac{B_0 N(N+1)}{k_B T}\right)}{\sum (2N+1) \exp\left(-\frac{BN(N+1)}{k_B T}\right)}$$

N: rotational quantum number, B_0: rotational constant

$$\langle E_{vib} \rangle_{ave} = \frac{hv}{2} + \frac{\sum n_v hv \exp\left(-\frac{n_v hv}{k_B T}\right)}{\sum \exp\left(-\frac{n_v hv}{k_B T}\right)} = \frac{hv}{2} + \frac{hv}{\exp\left(\frac{hv}{k_B T}\right) - 1}$$

n_v: vibrational quantum number, v: vibrational frequency   (6.5.3)

and the specific heat should be obtained from eq. (6.5.1).

In the solid state, the energy is given by the three-dimensional vibrational motion with neighboring mass (atoms or molecules). Einstein considered the independent harmonic vibrational motion of each mass for a single vibrational frequency [3]. Chapter 4.3 shows that the energy gap between neighboring harmonic vibrational states is $h\nu$ (ν: vibrational frequency). The average energy for three-dimensional harmonic vibration is given by:

$$\langle E \rangle_{ave} = \frac{3h\nu}{2} + \frac{3h\nu}{\exp\left(\frac{h\nu}{k_B T}\right)-1} \tag{6.5.3}$$

and the specific heat is:

$$c_v = \frac{3(h\nu)^2 \exp\left(\frac{h\nu}{k_B T}\right)}{k_B T^2 \left[\exp\left(\frac{h\nu}{k_B T}\right)-1\right]^2} \tag{6.5.4}$$

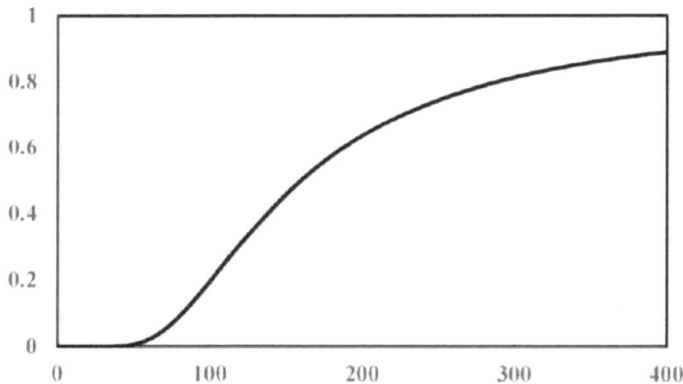

Fig. (6.3). Specific heat c_v (normalized by $3k_B$) of a solid obtained using eq. (6.5.4) as a function of the temperature T, using Einstein's model. The vibrational frequency was assumed to be 10^{13} Hz.

Fig. (**6.3**) shows the dependence of $c_v/(3k_B)$ as a function of T taking $\nu = 10^{13}$ Hz using Einstein's model. In this case, $c_v \to 3k_B$ for $T \to \infty$ and $c_v \to 0$ for $T \to 0$ were obtained. However, Einstein's model encounters a problem when describing c_v at low temperatures because it converges to zero as $T \to 0$ faster than the experimental result, which shows $c_v \propto T^3$. Debye considered an approach to treat the vibrational model as vibrations in a three-dimensional resonator considering the size of the object [4] and obtained a formula for the specific heat that was more consistent with experimental results compared to Einstein's model.

6.6.   POPULATION DISTRIBUTION WHEN EVENTS ARE NOT INDEPENDENT

The Boltzmann distribution shown in chapter 6.2 is derived from the probability distribution assuming the independent events. This assumption is not valid when atoms are cooled to ultra-low temperatures with high densities. Atomic waves are broadened larger than the interatomic distance, and there is interference between overlapped atomic waves. Thus, the quantum state cannot be an independent event. Before considering the distribution of the individual particles (atoms or molecules) at each state, we should introduce the fundamental characteristics of matter. When there are two identical particles 1 and 2, with the same quantum states at the a and b positions, the $[\varphi_1(a), \varphi_2(b)]$ and $[\varphi_1(b), \varphi_2(a)]$ states are not distinguishable, and there is an interference between the two states. For particles with spin S_p, the interference of both states is given by

$$\frac{\varphi_1(a)\ \varphi_2(b)+(-1)^{2S_p}\varphi_1(b),\varphi_2(a)}{\sqrt{2}} \tag{6.6.1}$$

Particles with integer spin are called Bosons, wherein the amplitude is larger by a factor of $\sqrt{2}$ (probability is two times larger) when $a = b$ (the position is equal to the distance that is smaller than the position uncertainty). For the interference between particles, Bosons tend to have the same quantum states. For particles with half-integer spin called Fermions, the wave function for $a = b$ is zero, and only one particle can exist in a single quantum state.

As shown in Fig. (**6.2**). the population in an energy state ϵ_E is proportional to $\Omega(\epsilon_E) \exp\left(-\frac{\epsilon_E}{k_B T}\right)$. This relationship also holds considering,

$$\Omega(\epsilon_E) \exp\left(-\frac{\epsilon_E}{k_B T}\right) \rightarrow \Omega(\epsilon_E) \exp\left(-\frac{\epsilon_E+\mu_C}{k_B T}\right) \tag{6.6.2}$$

where μ_C is virtual energy called the "chemical potential." Here, we consider the distribution of the number of particles in each state n_m. For Bosons, n_m can all be integers. For a given temperature, the average of n_m at each energy state is given by

$$\langle n_m(\epsilon_E)\rangle_{ave} = \frac{\Omega(\epsilon_E) \sum_{n=0}^{\infty} n_m \exp\left[-\frac{n_m(\epsilon_E+\mu_C)}{k_B T}\right]}{\sum_{n=0}^{\infty} \exp\left[-\frac{n_m(\epsilon_E+\mu_C)}{k_B T}\right]} = \frac{\Omega(\epsilon_E)}{\exp\left[\frac{(\epsilon_E+\mu_C)}{k_B T}\right]-1} \tag{6.6.3}$$

Where the chemical potential is given by the constant total matter number N_m as follows:

$$N_m = \int \frac{\Omega(\epsilon_E)}{\exp\left[\frac{(\epsilon_E + \mu_C)}{k_B T}\right] - 1} d\epsilon_E \tag{6.6.4}$$

Here, $\mu_C \geq 0$ is required so that $\langle n_m(0) \rangle_{ave} > 0$ is satisfied in addition to $\epsilon_E \to 0$. However, eq. (6.6. 4) is not always valid. For an ultra-low temperature,

$$N_{m0} = \int \frac{\Omega(\epsilon_E)}{\exp\left[\frac{(\epsilon_E + \mu_C)}{k_B T}\right] - 1} d\epsilon_E < \int \frac{\Omega(\epsilon_E)}{\exp\left[\frac{\epsilon_E}{k_B T}\right] - 1} d\epsilon_E \approx \int \Omega(\epsilon_E) \exp\left[-\frac{\epsilon_E}{k_B T}\right] < N_m \tag{6.6.5}$$

Bosons with a number $(N_m - N_{m0})$ are condensate to the lowest energy (see Fig.6-7). This phenomenon is called Bose-Einstein condensation (BEC). The phase of the matter waves in the BEC is uniform, and the group treats a single mechanical entity using a wave function on a macroscopic state. Here, we discuss the condition for obtaining the BEC state in free space. For a given momentum p_q ($q = x, y, z$), the number of states of the position in the region $0 < q < L$ is $\frac{L p_q}{h} = \frac{L}{\lambda}$, where λ is the wavelength of the matter wave. The number of states with the momentum between p_q and $p_q + dp_q$ is given by $\Omega(p_q) dp_q = \frac{L}{h}$. Considering the state of number with the three-dimensional kinetic energy,

$$\Omega(p) dp = \Omega(p_x)\Omega(p_y)\Omega(p_z) dp_x dp_y dp_z = \left(\frac{L}{h}\right)^3 (4\pi p^2) dp$$

$$\Omega(\epsilon_E) d\epsilon_E = \frac{4\pi}{h^3} V (2m)^{\frac{3}{2}} \sqrt{\epsilon_E} d\epsilon_E$$

m; mass V: volume

$$N_{m0} < \int \frac{\Omega(\epsilon_E)}{\exp\left[\frac{\epsilon_E}{k_B T}\right] - 1} d\epsilon_E = \frac{4\pi}{h^3} V (2mk_B T)^{\frac{3}{2}} \int \frac{\sqrt{x}}{\exp[x] - 1} dx \tag{6.6.6}$$

Using

$$\int \frac{\sqrt{x}}{\exp(x) - 1} dx = \frac{\sqrt{\pi}}{2} \times 2.612 \tag{6.6.7}$$

the BEC state ($N_m > N_{m0}$) is obtained when [5,6]:

$$\frac{N_m}{V}\left(\frac{h^2}{4\pi m k_B T}\right)^{\frac{3}{2}} > 2.612 \qquad\qquad (6.6.8)$$

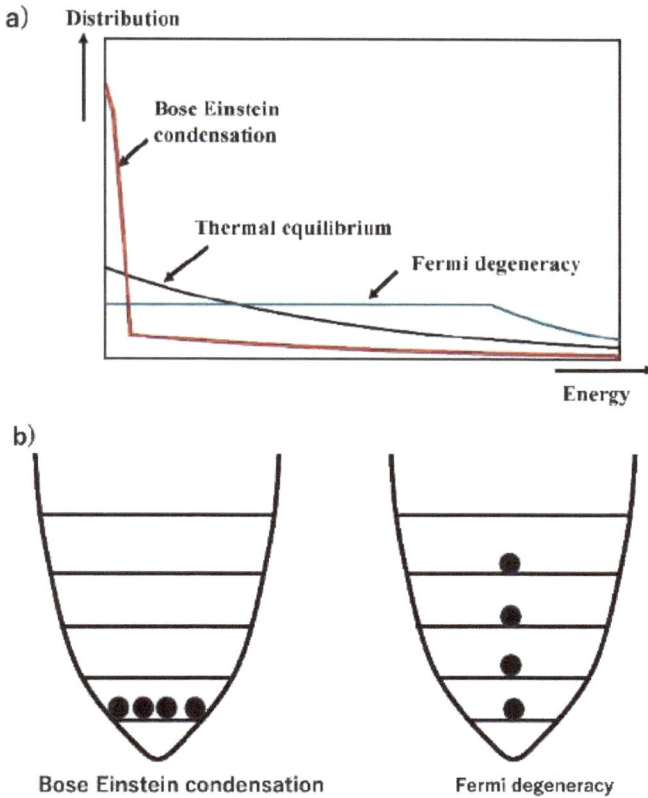

Fig. (6.4). (a) Energy distribution using thermal equlibrium, Bose Einstein condensation, and Fermi degeneracy. (b) Energy distribution of four atoms in Bose Einstein condensation and Fermi degeneracy. (This figure is used also in "Cold Atoms and Molecules" by M. Kajita).

For Fermion particles, n_m can only be 0 or 1, and the average of n_m at each energy state is given by:

$$\langle n_m(\epsilon_E)\rangle_{ave} = \frac{\Omega(\epsilon_E)\exp\left[-\frac{\epsilon_E+\mu_C}{k_BT}\right]}{1+\exp\left[-\frac{\epsilon_E+\mu_C}{k_BT}\right]} = \frac{\Omega(\epsilon_E)}{\exp\left[\frac{(\epsilon_E+\mu_C)}{k_BT}\right]+1}$$

$$N_m = \int \frac{\Omega(\epsilon_E)}{\exp\left[\frac{(\epsilon_E+\mu_C)}{k_BT}\right]+1}\,d\epsilon_E \qquad (6.6.9)$$

Here, the chemical potential is negative and for $T \to 0$,

$$\langle n_m(\epsilon_E)\rangle_{ave} = \Omega(\epsilon_E) \qquad \epsilon_E + \mu_C < 0$$

$$\langle n_m(\epsilon_E)\rangle_{ave} = 0 \qquad \epsilon_E + \mu_C > 0 \qquad (6.6.10)$$

For a group of Fermions, energy distribution cannot be localized in the lowest state at ultra-low temperatures, but they can occupy the states in order from the lowest, as shown in Fig. (**6.4**). This is called "Fermi degeneracy". The Fermi energy $\epsilon_F(= -\mu_C)$ is defined as a parameter of energy to show energy distribution broadening when $T = 0$. The Boltzmann distribution is valid when $k_BT \gg \epsilon_F$. The simplest example is an electron in an atom; only one electron can assume the state given by principal quantum number n, rotational quantum number L, magnetic quantum number M, and electron spin quantum number in the z-direction S_z (see chapter 5.1). In the 1S state, only two electrons are present. For the Li atom, the third electron must be in the $n = 2$ state.

However, Fermions can pair up with opposite spins and have Boson like states (Cooper pair). Paired Fermions can also undergo the BEC state. BECs are related to the superfluidity (^{3}He and ^{4}He form a liquid that flows with zero friction) and superconductivity (BEC state of Cooper paired electrons results in zero electric resistance). The motion of all particles in the BEC state is macroscopic, and any scattering of a small fraction of particles is forbidden.

6.7. THE IDEAL GAS LAW AND THE WORK DONE BY THE GAS PRESSURE

For an ideal gas (volume of molecules and the intermolecular interaction are ignored), the relationship between the pressure P, volume V, number of gaseous molecules N_m, and temperature T is given by:

$$PV = N_m k_B T \qquad \text{(called "ideal gas law")} \qquad (6.7.1)$$

which is derived by considering the collision between the gas and wall. Considering the mass and velocity of the gas molecules m and v, respectively, the change of the momentum for a single collision is given by (see Fig. **6.5**)

$$\Delta p = 2mv \tag{6.7.2}$$

velocity change: $v \to -v$

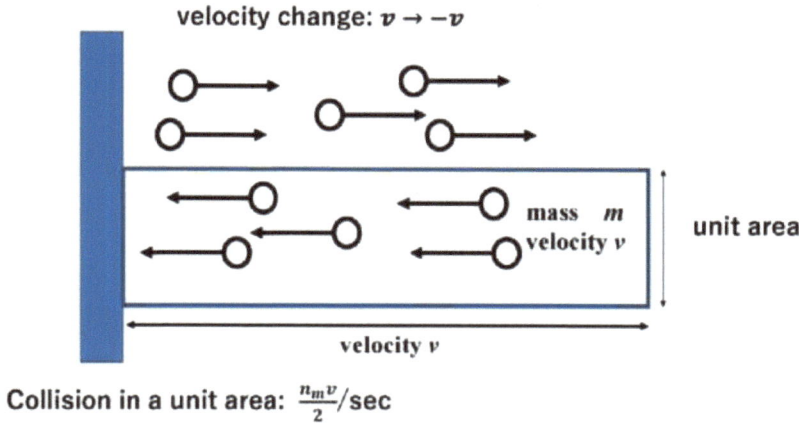

mass m
velocity v

unit area

velocity v

Collision in a unit area: $\frac{n_m v}{2}$/sec

Fig. (6.5). Collision between the gaseous molecule (mass m, velocity v). The momentum change after a single collision is $2mv$ and the collision on a unit area on the wall occurs $\frac{n_m v}{2}$ times per unit time (n_m: molecular density).

On a unit area of the wall, the number of collisions with the gas molecules per unit time is given by:

$$\Gamma_{col-wall} = \frac{n_m v}{2} \quad n_m = \frac{N_m}{V}$$

(half of the molecules have a motion to the wall) $\tag{6.7.3}$

The pressure is then given by:

$$P = \Delta p \times \Gamma_{col-wall} = mv^2 \frac{N_m}{V} \tag{6.7.4}$$

Considering eq. (6.3.4), the average of mv^2 is $k_B T$ and eq. (6.7.1) can be derived.

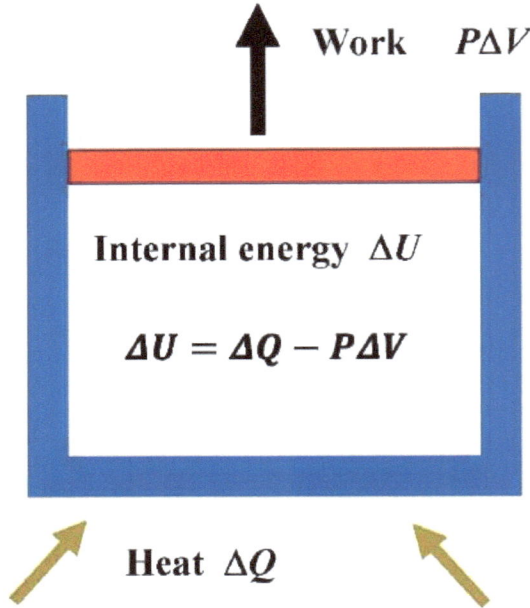

Fig. (6.6). The relationship between the change of the gas internal energy ΔU, heat from outside ΔQ, and work $P\Delta V$.

When the volume of the gas changes by ΔV, the system performs work given by $P\Delta V$. For the absorption of heat ΔQ (flow of energy from the matter at a higher temperature), the change in the internal energy (kinetic energy of the center of mass, molecular rotational, or vibrational energy) ΔU is given as (see Fig. **6.6**):

$$\Delta U = \Delta Q - P\Delta V \tag{6.7.5}$$

Using the specific heat at constant volume c_v, which was discussed in chapter 6.5,

$$\Delta U = N_m c_v \Delta T \tag{6.7.6}$$

and the general specific heat is given by

$$c_g = \frac{1}{N_m}\frac{\Delta Q}{\Delta T} = c_v + \frac{P}{N_m}\frac{\Delta V}{\Delta T} \tag{6.7.7}$$

The specific heat at constant pressure is given by

$$\Delta V = \frac{N_m}{P} k_B \Delta T \qquad c_p = c_v + k_B \tag{6.7.8}$$

For adiabatic expansion or compression ($\Delta Q = 0$), we have:

$$\Delta U = N_m c_v \Delta T = -P\Delta V \tag{6.7.9}$$

For expansion (compression), the temperature becomes lower (higher), as shown in eq. (6.7.9). Using eqs. (6.7.1) and (6.7.9), we have:

$$N_m c_v \Delta T = -\frac{N_m k_B T}{V} \Delta V$$
$$\frac{c_v}{T} \Delta T = -\frac{k_B}{V} \Delta V$$
$$\ln(T) = -\frac{k_B}{c_v} \ln(V) + C \quad TV^{\frac{k_B}{c_v}} = const \tag{6.7.10}$$

For rare gas atoms (He, Ne, Ar, Kr, Xe), $c_v = \frac{3}{2} k_B$ (see chapter 6.5) and $TV^{\frac{2}{3}}$ is constant.

The Carnot cycle is a theoretical ideal thermodynamic cycle for the absorption of heat and performing the following four procedures [7].

(1) Absorption of the heat of $Q_1 > 0$ and expansion of the volume of gas ($V = V_1 \rightarrow V_2$) at a constant temperature T_1.
(2) Adiabatic expansion $V = V_2 \rightarrow V_3 (> V_2)$ and $T = T_1 \rightarrow T_2 (< T_1)$.
(3) Absorption of the heat of $Q_2 < 0$ and the compression of the volume of gas ($V = V_3 \rightarrow V_4$) at a constant temperature T_2.
(4) Adiabatic compression $V = V_4 \rightarrow V_1 (< V_4)$ and $T = T_2 \rightarrow T_1$.
The work done for each procedure is

$$W_1 = \int_{V_2}^{V_2} PdV = N_m k_B T_1 \ln\left(\frac{V_2}{V_1}\right) > 0$$
$$W_2 = N_m c_v (T_1 - T_2) > 0$$
$$W_3 = N_m k_B T_2 \ln\left(\frac{V_4}{V_3}\right) < 0$$
$$W_4 = N_m c_v (T_2 - T_1) < 0 \tag{6.7.11}$$

Equation (6.7.10) shows that:

$$T_1 V_2^{\frac{k_B}{c_v}} = T_2 V_3^{\frac{k_B}{c_v}} \quad T_1 V_1^{\frac{k_B}{c_v}} = T_2 V_4^{\frac{k_B}{c_v}} \quad \frac{V_2}{V_1} = \frac{V_3}{V_4} \qquad (6.7.12)$$

and the total work done in a cycle is obtained using $W_3 = -\frac{T_2}{T_1}$ and $W_2 + W_4 = 0$ as:

$$W_{cycle} = W_1 + W_2 + W_3 + W_4 = N_m k_B (T_1 - T_2) \ln\left(\frac{V_2}{V_1}\right) \qquad (6.7.13)$$

For the ideal model,

$$Q_1 = W_1 \quad Q_3 = W_3 \qquad (6.7.14)$$

and the work efficiency (ratio of the work to the absorbed heat) is

$$\eta_W = \frac{W_{cicle}}{Q_1} = \frac{T_1 - T_2}{T_1} \qquad (6.6.15)$$

Here, the thermodynamical entropy is defined by:

$$\Delta S = \frac{\Delta Q}{T} \qquad (6.7.16)$$

With eqs.(6.7.11) and (6.7.14), the change in entropy is

$$\Delta S = \frac{Q_1}{T_1} + \frac{Q_2}{T_2} = \frac{W_1}{T_1} + \frac{W_3}{T_2} = 0 \qquad (6.7.17)$$

Using entropy, the work in the Carnot cycle is given by:

$$W_{cycle} = \oint T dS \qquad (6.7.18)$$

which is given by the area surrounded by the square shown in Fig. (**6.7**).

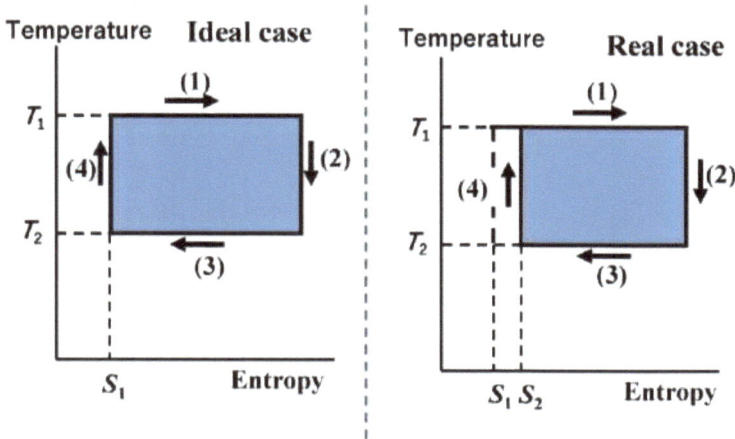

Fig. (6.7). The relationship between entropy and temperature in the Carnot cycle. The area surrounded by the square denotes the work done during the cycle. For an ideal case, the entropy returns to the initial value after the cycle. However, in practice, the entropy becomes higher than the initial value and the area becomes smaller than the ideal case.

In practice, there is energy loss due to friction and Eqs. (6.7.14-15) are corrected to:

$$Q_1 > W_1 \quad Q_3 > W_3 \tag{6.7.19}$$

$$\eta_W = \frac{W_{cicle}}{Q_1} < \frac{T_1 - T_2}{T_1} \tag{6.7.20}$$

Then eq. (6.7.17) is corrected to:

$$\Delta S > 0 \tag{6.7.21}$$

This means that the entropy increases for each cycle. As shown in Fig. (6.7), the entropy after one cycle S_2 is higher than the initial value S_1, and the area that represents the work done becomes smaller than that of the ideal model.

Equation (6.7.21) describes the flow of heat between masses 1 and 2 with temperatures T_1 and T_2 ($T_1 > T_2$). Then the heat of $Q > 0$ flows from mass 1 to mass 2, and the change of the entropy is given by:

$$\Delta S = -\frac{Q}{T_1} + \frac{Q}{T_2} > 0 \tag{6.7.22}$$

In statistical mechanics, the temperature is defined using statistical mechanical entropy. Rewriting eq. (6.2.6), we have:

$$\frac{1}{T} = \frac{dS}{dE} \qquad dS = \frac{dE}{T} \tag{6.7.23}$$

which is equivalent to eq. (6.7.16). Therefore, the thermodynamic entropy is the same as the statistical mechanical entropy, except that entropy is defined using temperature.

The fundamentals of statistical mechanics and thermodynamics are that entropy is always increasing, and a state is in equilibrium when the entropy is maximum (the probability is a maximum).

Note that entropy is divided into that which is inside (S_{in}) and outside (S_{out}) of the system and:

$$\Delta S_{in} + \Delta S_{out} \geq 0 \tag{6.7.24}$$

holds. Using $T\Delta S_{out} = \Delta Q_{out} = -\Delta Q_{in}$ and $\Delta Q_{in} = \Delta U + P\Delta V$, eq. (6.7.24) is rewritten as:

$$\Delta U + P\Delta V - T\Delta S_{in} \leq 0 \tag{6.7.25}$$

For a constant temperature and constant volume, eq. (6.7.25) is equivalent to

$$\Delta F \leq 0 \qquad F = U - TS \tag{6.7.26}$$

where F is called the "Helmholtz's free energy". The thermal equilibrium state is attained with a minimum value of F. The derivative form of F is given as:

$$\Delta F = \Delta U - T\Delta S - S\Delta T$$

using $T\Delta S = \Delta U + P\Delta V$

$$\Delta F = -P\Delta V - S\Delta T \leq 0 \tag{6.7.27}$$

and the thermal equilibrium state is given by:

$$\Delta F = 0, \qquad \frac{\Delta V}{\Delta T} = -\frac{S}{P} \tag{6.7.28}$$

At a constant temperature and constant pressure, eq. (6.7.25) is equivalent to:

$$\Delta G \leq 0 \qquad G = U + PV - TS \tag{6.7.29}$$

where G is the "Gibbs's free energy". The thermal equilibrium state is attained with the minimum value of G. The derivative form of G is given as:

$$\Delta G = V\Delta P - S\Delta T \leq 0 \tag{6.7.30}$$

and the thermal equilibrium state is given by:

$$\Delta G = 0, \qquad \frac{\Delta P}{\Delta T} = -\frac{S}{V} \tag{6.7.31}$$

The dependence of the vapor pressure of matter on temperature is obtained as follows using $V = \frac{N_m k_B T}{P}$ and $S = \frac{N_m L_{ev}}{T}$ (L_{ev}: evaporation energy);

$$\frac{1}{P}\frac{\Delta P}{\Delta T} = -\frac{L_{ev}}{k_B T^2} \qquad P \propto \exp\left(-\frac{L_{ev}}{k_B T}\right) \tag{6.6.32}$$

Equation (6.6.31) can also be derived from the Boltzmann distribution of the molecules in the gas phase for energy higher than that of the liquid- or solid-state by L_{ev}.

EXERCISE

(1) For the rotational quantum number N, number of states $2N + 1$ and energy $B_0 N(N + 1)$ B_0: rotational constant

Obtain the rotational state for which the population has a maximum value.
(Answer)

The population is proportional to $P(N) = (2N + 1)\exp\left(-\frac{B_0 N(N+1)}{k_B T}\right)$

For the maximum population, $\frac{dP(N)}{dN} = 0$

Thus, we have a maximum when $N = \dfrac{\sqrt{\dfrac{2k_B T}{B_0}} - 1}{2}$

(2) For the cooling of He gas from T to $T/2$ by adiabatic expansion, the volume must be expanded by $V \rightarrow V'$. Obtain the value of V'/V.

(Answer)

Equation (6.7.10) shows the constancy of $TV^{\frac{k_B}{c_v}}$. For He, $c_v = \frac{3}{2}k_B$, as shown in Eq. (6.5.2). The answer is $2\sqrt{2}$.

REFERENCES

[1]   Value: Boltzmann constant (nist.gov),

[2]   https://www.higherpeak.com/altitudechart.html

[3]   A. Einstein, "Die Plancksche Theorie der Strahlung und die Theorie der spezifischen Wärme", *Ann. Phys.,* vol. 327, no. 1, pp. 180-190, 1906.
        [http://dx.doi.org/10.1002/andp.19063270110]

[4]   P. Debye, "Zur Theorie der spezifischen Waerme", *Ann. Phys.,* vol. 344, no. 14, pp. 789-839, 1912.
        [http://dx.doi.org/10.1002/andp.19123441404]

[5]   M.H. Anderson, J.R. Ensher, M.R. Matthews, C.E. Wieman, and E.A. Cornell, "Observation of bose-einstein condensation in a dilute atomic vapor", *Science,* vol. 269, no. 5221, pp. 198-201, 1995.
        [http://dx.doi.org/10.1126/science.269.5221.198] [PMID: 17789847]

[6]   D.M. Stamper-Kurn, M.R. Andrews, A.P. Chikkatur, S. Inouye, H-J. Miesner, J. Stenger, and W. Ketterle, "Optical confinement of a bose-einstein condensate", *Phys. Rev. Lett.,* vol. 80, no. 10, pp. 2027-2030, 1998.
        [http://dx.doi.org/10.1103/PhysRevLett.80.2027]

[7]   I.A. Martínez, É. Roldán, L. Dinis, *et al.*, "Brownian carnot engine", *Nat. Phys.,* vol. 12, no. 1, pp. 67-70, 2016. [http://dx.doi.org/10.1038/nphys3518] [PMID: 27330541]

CHAPTER 7

Analysis of the Measurement Uncertainties

Abstract: Physical laws, which have finite uncertainties, are established to facilitate measurements. New physical phenomena have been discovered when the measurement uncertainties were reduced. The object of this chapter is to review the estimation of the measurement uncertainties. This parameter consists of statistical uncertainty and systematic uncertainty. Statistical uncertainty is given by the random distribution of measurement results in a region with a broadening of σ in the vicinity of a real value. The statistical uncertainty of the average of the measurement results is expected to be $\frac{\sigma}{\sqrt{N}}$, where N is the number of measurement samples. Systematic uncertainty exists because measurements are influenced by the conditions under which they are obtained. The real value is defined for a certain circumstance, and the measurements obtained under different circumstances are shifted from this value. The real value is obtained by correcting the shift, which is estimated according to the measurement circumstances. Statistical uncertainty is obtained from the uncertainty in the estimation of the shifts.

Keywords: Atomic clock, Central limit theorem, CPT-symmetry, Gravitational potential in the micro-scale, Measurement uncertainty, Ptolemaic and Copernican systems, Spectrum broadening, Stark shift, Statistic uncertainty, Systematic uncertainty, Zeeman shift.

7.1. IMPORTANCE OF THE MEASUREMENT UNCERTAINTY [1]

Physics involves the study of the laws of nature, from which we can make predictions regarding future phenomena. This is based on measurement results. For example, for the measurement results of $(x,y) = (1,1), (2,2), (3,3), (5,5)$, the law of $y = x$ is established. However, uncertainty is always present in measurement results. If the measurement uncertainty is 10%, the proportionality cannot be established. The measurement result for $x > 20$ may have a discrepancy with the established law $y = x$ (Fig. **7.1**).

Physical laws were established based on the measurement results obtained at a given time. The development of new physical laws is necessary to account for the discovery of new phenomena that are not consistent with previous physical laws. New phenomena have often been discovered *via* the minimization of measurement uncertainties. If the uncertainty of measurements are reduced to 0.5% and $(x,y) =$

(1,0.99), (2,1.98), (3,2.95), (5,4.79) are obtained, $y = 10\sin(x/10)$ is more appropriate description than $y = x$.

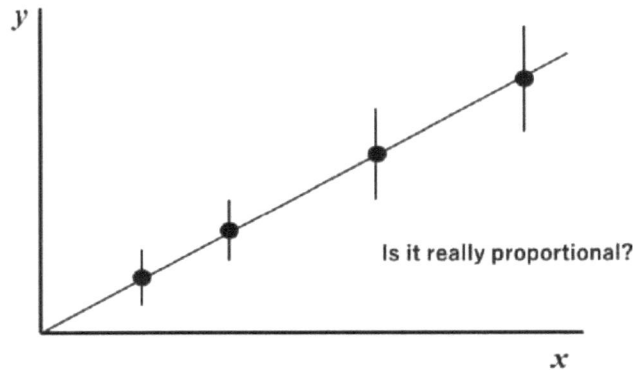

Fig. (7.1). All measurements have some degree of uncertainty; therefore, laws from measurement results should always be subjected. (This figure is used also in "Measurement, Uncertainty and Lasers" by M. Kajita).

The development of physics has been closely correlated to improving the accuracy of time and frequency, especially with respect to the invention of new clocks [2]. In ancient times, the Ptolemaic system was accepted because there was no discrepancy related to measurement uncertainties when it was adopted. As the accuracy of the clock improved, discrepancies were discovered regarding the position of stars at certain moments, and the Copernican system was established. Newtonian mechanics was established within 100 years after the accuracy of the clock was drastically improved by the discovery of the periodicity of the pendulum. As the accuracy of clocks continued to improve, the fluctuation of the orbital period of Io (satellite of Jupiter) was observed to be out of the margin of error of these instruments. To interpret this phenomenon, the speed of light was recognized as being finite for the first time; the light propagation time is not constant because the distance between the earth and Jupiter changes owing to their orbital motions. The characteristics of light were further elucidated because the propagation of electromagnetic waves, as described by Maxwell's equation, is in good agreement with the measured speed of light.

Measurements with ultra-low uncertainty have been performed for atomic transition frequencies. Atomic clocks are based on the atomic transition frequencies. Fractional measurement uncertainties below 10^{-17} have been obtained for several transition frequencies [3-7].

The slight shift in the transition frequencies significantly contributed to the development of modern physics. For example, time slows down in a moving frame, as shown in chapter 3.6. However, for a velocity of 1000 m/s, this effect results in a fractional shift of 10^{-11}, which can only be detected using atomic clocks. The theory of general relativity shows that time shows down in the presence of a strong gravitational field (gravitational redshift). This effect results in a fractional shift of 10^{-16} with a change in altitude of 1 m. For a measurement uncertainty below 10^{-17}, this effect is detected when there is a change in altitude of 15 cm [3].

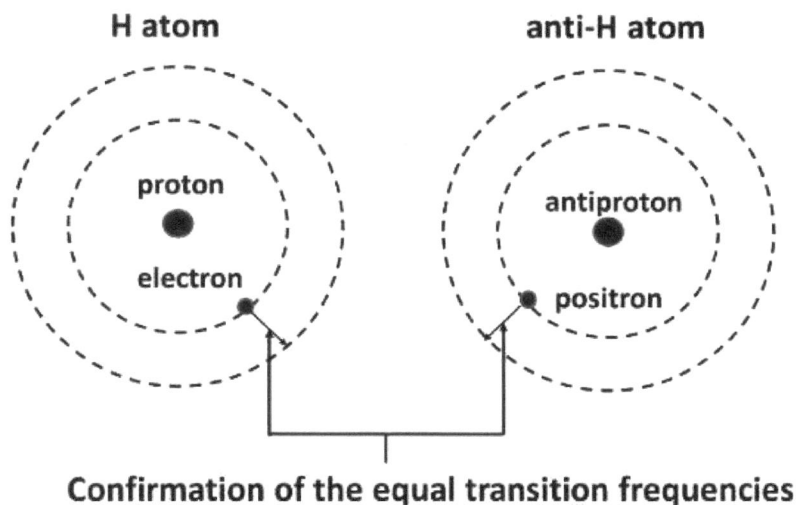

Confirmation of the equal transition frequencies

Fig. (7.2). Comparison of transition frequencies of an H-atom and an anti-H atom. (This figure is used also in "Measurement, Uncertainty and Lasers" by M. Kajita).

Further developments in physics are expected based on the detection of slight effects, for which measurements with ultralow uncertainties are required. For example, there is a mystery concerning the relationship between particles and antiparticles. Antiparticles are expected to be charge-conjugated and mirror-image inverted particles (CP-symmetry). However, why aren't antiparticles present in nature? If the CP-symmetry is violated, the number of particles can be larger with a ratio of 10^{-9} compared to antiparticles, and particles can continue to exist after their antiparticles disappear *via* pair annihilation. A violation of CP-symmetry was discovered in 1964 [8]. However, CPT symmetry (antiparticles must have the image of particles after charge-conjugation, mirror-image inversion, and time-reversal) is required to maintain the Lorenz invariance; all physical laws in a coordinate must also hold after the Lorentzian transform. CTP-symmetry was confirmed within the current measurement uncertainties. For example, the equality of the 1S-2S transition frequencies of hydrogen and anti-hydrogen atoms was confirmed with an uncertainty of 2×10^{-12} [9] (see Fig. **7. 2**).

The Coulomb and gravitational potential energies were determined to be proportional to $\frac{1}{r}$ at the macroscopic scale, where r is the distance between two bodies. For the Coulomb potential energy, this law also holds at the microscopic scale. However, gravitational potential energy is not guaranteed at the microscopic scale. Suppose the formula for the gravitational potential energy at the micro scale is different from that at the macro scale. In that case, it might be useful to unify the general theory of relativity and quantum mechanics, which has not been achieved to date. However, the gravitational potential term on the microscale is much less than the electromagnetic potential, and the measurement of the interatomic potential (vibrational transition frequency of molecule) with an ultra-low uncertainty is required [10].

This chapter introduces the fundamentals that are required to estimate measurement uncertainties.

7.2. STATISTICAL AND SYSTEMATIC UNCERTAINTIES [11]

When performing a physical measurement, there is a question of whether it is a real value. The reliability of measured values can be confirmed by repeating the measurements many times. The measurement results are expected to be distributed over a certain limited range. The uncertainty given by the non-zero distribution area is called the "statistical uncertainty", as shown in Fig. **(7.3)**.

The measurement results might be distributed in an area shifted from the real value because the measurement values can vary according to a given set of circumstances. The real values should be defined under a specified condition, and the measured value is shifted with another circumstance. If we know the dependence of the measurements on the circumstances, we can correct the measured values using the estimated shift. Then the uncertainty of the estimated shift becomes another uncertainty, called the "systematic uncertainty", as shown in Fig. **(7.3)**.

More detailed explanations of statistical and systematic uncertainties are presented in the following subsections.

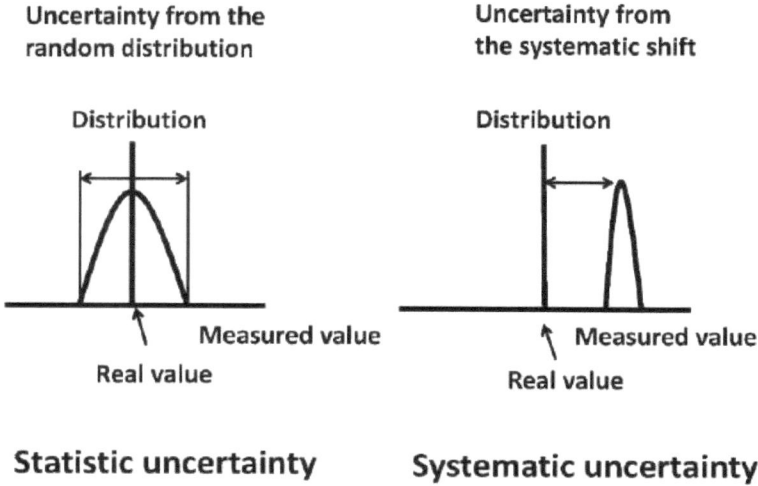

Fig. (7.3). Statistic and systematic uncertainties with the distributions of measurements. (This figure is used also in "Measurement, Uncertainty and Lasers" by M. Kajita).

7.2.1 Statistical Uncertainty

Statistical uncertainty is given by the finite broadening of the distribution area of the measurement results. This broadening can be induced by the temporal fluctuation of the event, which can be reduced by stabilizing the circumstance. This broadening is also the result of the quantum uncertainty principle.

We considered the measurement of the physical value X. The measurement results are distributed around the real value for the measurement condition X_r. Taking measurement samples $X(i)$ ($i = 1 - N$), we obtain the average X_{ave} and the broadening of the distribution range σ. When N is sufficiently large, the probability distribution of X_{ave} is approximately given by (Fig. **7.4**).

$$P(X_{ave}) = \frac{\sqrt{N}}{\sigma\sqrt{\pi}} \exp\left[-N\left(\frac{X_{ave}-X_r}{\sigma}\right)^2\right] \qquad (7.2.1)$$

for any distribution of measurement results (central limit theorem). Therefore, X_r is estimated to be in the range given by

$$X_r = X_{ave} \pm \frac{\sigma}{\sqrt{N}} \tag{7.2.2}$$

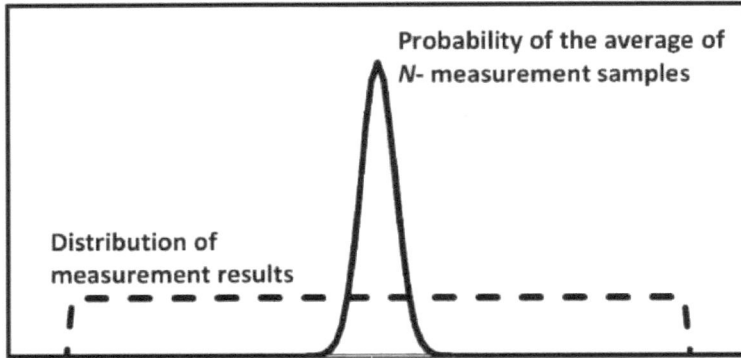

Fig. (7.4). Relationship between the distributions of the measurement results and the probability of the average of N-measurement samples. The probability of the average of the N-measurement samples is given by a Gaussian distribution with a broadening that is narrower than that of the measurement results by a factor of $1/\sqrt{N}$. (This figure is used also in "Measurement, Uncertainty and Lasers" by M. Kajita).

Statistical uncertainty is reduced by increasing the number of measurement samples. It is not simple to derive the central limit theorem as a general formula [12]. Here, we consider the average of the measurement results when we obtain $X_r + \sigma$ and $X_r - \sigma$ with a probability of 1/2. Repeating the measurement N times, the probability $p(n)$ to measure n times $X_0 + \sigma$ and $(N - n)$ times $X_0 - \sigma$ is given by:

$$p(n) = \frac{N!}{n!(N-n)!} \left(\frac{1}{2}\right)^N \tag{7.2.3}$$

Here, we consider the Taylor expansion of $\ln[p(n)]$ using:

$$\frac{d\ln[p(n)]}{dn} = -\ln(n) + \ln(N-n), \frac{d^2\ln[p(n)]}{dn^2} = -\frac{1}{n} - \frac{1}{N-n} \tag{7.2.4}$$

Here, $\ln[p(n)]$ at n close to $N/2$ is approximately given by:

$$\ln[p(n)] = \ln p\left(\frac{N}{2}\right) - \frac{4}{N}\left(n - \frac{N}{2}\right)^2,$$
$$p(n) = p(0)\exp\left[-\frac{4}{N}\left(n - \frac{N}{2}\right)^2\right] \tag{7.2.5}$$

The average of the measurement is given by:

$$X_{ave} = X_r + \frac{2n-N}{N}\sigma,$$

$$n - \frac{N}{2} = \frac{N}{2\sigma}(X_{ave} - X_r) \tag{7.2.6}$$

Then the probability distribution of X_{ave} is given by:

$$p(X_{ave}) = p(X_r)\exp\left[-N\left(\frac{X_{ave}-X_r}{\sigma}\right)^2\right] \tag{7.2.7}$$

The measurement of the transition frequencies of atoms and molecules ν_0 are distributed within a frequency area of $\nu_0 - \gamma < \nu < \nu_0 + \gamma$, where γ is the spectrum broadening given by the limited time of the interaction between the atoms or molecules and light in the absence of a phase jump (see chapter 4.6). The statistical measurement uncertainty is given by:

$$(\Delta\nu)_{stat} = \frac{\gamma}{\nu_0}\sqrt{\frac{\tau_m}{N_a T_m}} \tag{7.2.8}$$

where τ_m is a time for the single measurement, T_m is the measurement time, and N_a is the number of atoms or molecules. Note $\tau_m > \frac{1}{2\pi\gamma}$ and $(\Delta\nu)_{stat} \propto \sqrt{\gamma}$ when $\tau_m \approx \frac{1}{2\pi\gamma}$.

7.2.2. Systematic Uncertainty

Statistical uncertainty is reduced by averaging many measurement results, but it might not always converge to the real value. The measured values are dependent on the circumstances, and the real values should be defined for a specific circumstance. By measuring under a different circumstance, the measurement result is shifted from the defined value. Systematic uncertainty is reduced by monitoring the different circumstances and correcting the estimated shifts. For this correction, the systematic uncertainty is given by the uncertainty of the estimated measurement shift.

For example, thermal expansion causes a change in the length of an object; therefore, the real length should be defined at a specific temperature (T_0). By repeating the measurement with another temperature T_p, the average length is shifted from the defined length (Fig. **7.5**). Systematic uncertainty can be reduced by monitoring T_p and applying a correction $\alpha_p(T_0 - T_p)$; however, it cannot be zero because of the uncertainties of T_p and α_p (α_p is the coefficient of thermal expansion).

For the transition frequencies of atoms or molecules, the shifts of the measurements are induced by the electric field (Stark shift) and the magnetic field (Zeeman shift). These frequency shifts can be eliminated by the measurement or theoretical estimation of the Stark and Zeeman shifts. The measurements of the transition frequencies are also shifted owing to relativistic effects, as shown in chapter 7.1.

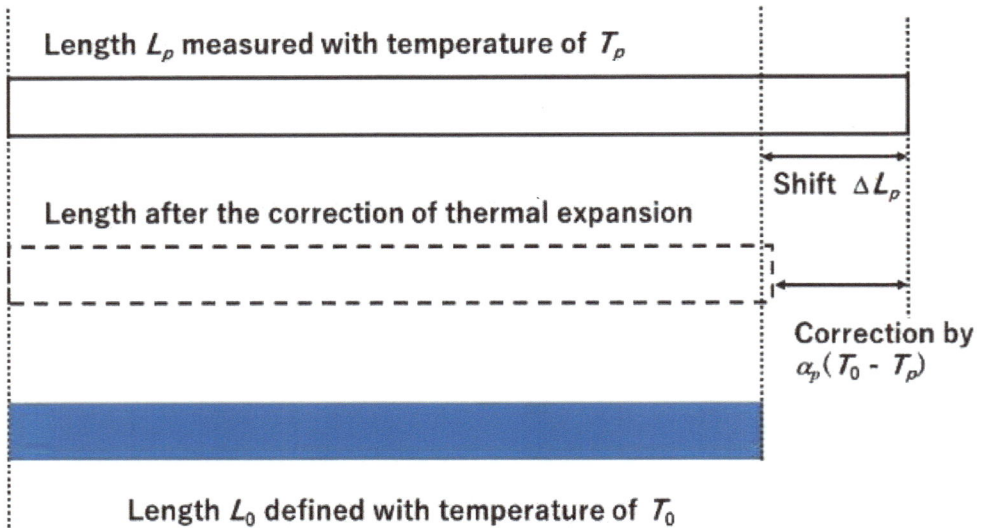

Fig. (7.5). Concept of systematic uncertainty associated with a length measurement. The length should be defined at a specific temperature T_0. There is a shift in the measurement at a different temperature T_p. The systematic uncertainty is reduced by monitoring the temperature and applying a correction of thermal expansion $\alpha_p(T_0 - T_p)$ (α_p: thermal expansion coefficient), which cannot be zero because of the uncertainties of T_p and α_p. (This figure is used also in "Measurement, Uncertainty and Lasers" by M. Kajita).

7.3. TOTAL UNCERTAINTY

As shown in chapter 7.2, there are statistical and systematic uncertainties, δX_{St} and δX_{Sy}.

There is no correlation between the two uncertainty terms, and the total measurement uncertainty can be estimated as:

$$\delta X_{tot} = \sqrt{(\delta X_{st})^2 + \left(\delta X_{sy}\right)^2} \tag{7.3.1}$$

which is a parameter that gives the total measurement uncertainty. Systematic uncertainty is the combination of the uncertainties of different shifts in the measurement results, which can be represented as

$$\delta X_{Sy} = \sqrt{\Sigma\left(\delta X_{Sy-i}\right)^2} \tag{7.3.2}$$

Using this estimation of the total measurement uncertainty, the total uncertainty is given by one or a few causes of uncertainties which are the most dominant.

EXERCISE

Assume that we get the measurement results $X_r + \sigma$ and $X_r - \sigma$ with the probability of 1/2. From the measurement samples of N, the probability $P(n)$ to get both results n and $N - n$ is given by Eq. (7.2.3). With $n = \frac{N}{2}$, $P(n)$ is maximum and the averaged value becomes X_r.

Obtain $\frac{P\left(\frac{N}{4}\right)}{P\left(\frac{N}{2}\right)}$ with $N = 4, 16$, and 64. You will see that the probability to have the averaged value of $X_{ave} = X_r \pm \frac{\sigma}{2}$ becomes smaller as N increases.
(Answer)

$N = 4$ 0.67 $N = 16$ 0.14 $N = 64$ 0.00027

REFERENCES

[1]   M. Kajita, *Measurement, Uncertainty and Lasers.* IOP Expanding Physics, 2019, pp. 1-17.
       [http://dx.doi.org/10.1088/2053-2563/ab0373]

[2]   M. Kajita, *Measuring Time; Frequency measurements and related developments in physics.* IOP
       Expanding Physics, 2018, pp. 1-12.

[3]   C-W. Chou, D.B. Hume, J.C.J. Koelemeij, D.J. Wineland, and T. Rosenband, "Frequency comparison
       of two high-accuracy Al+ optical clocks", *Phys. Rev. Lett.*, vol. 104, no. 7, p. 070802, 2010.
       [http://dx.doi.org/10.1103/PhysRevLett.104.070802] [PMID: 20366869]

[4] N. Huntemann, C. Sanner, and B. Lipphardt, "Chr. Tamm, and E. Peik. Single-Ion Atomic Clock with", *Phys. Rev. Lett.,* vol. 116, p. 063001, 2016.
[http://dx.doi.org/10.1103/PhysRevLett.116.063001] [PMID: 26918984]

[5] I. Ushijima, M. Takamoto, M. Das, T. Ohkubo, and H. Katori, "Cryogenic optical lattice clocks", *Nat. Photonics,* vol. 9, no. 3, pp. 185-189, 2015.
[http://dx.doi.org/10.1038/nphoton.2015.5]

[6] T.L. Nicholson, S.L. Campbell, R.B. Hutson, G.E. Marti, B.J. Bloom, R.L. McNally, W. Zhang, M.D. Barrett, M.S. Safronova, G.F. Strouse, W.L. Tew, and J. Ye, "Systematic evaluation of an atomic clock at $2 \times 10(-18)$ total uncertainty", *Nat. Commun.,* vol. 6, p. 6896, 2015.
[http://dx.doi.org/10.1038/ncomms7896] [PMID: 25898253]

[7] N. Nemitz, T. Ohkubo, M. Takamoto, I. Ushijima, M. Das, N. Ohmae, and H. Katori, "Frequency ratio of Yb and Sr clocks with 5", *Nat. Photonics,* vol. 10, no. 4, pp. 258-261, 2016.
[http://dx.doi.org/10.1038/nphoton.2016.20]

[8] J.H. Christenson, J.W. Cronin, V.L. Fitch, and R. Turlay, "Evidence for the 2π Decay of the K20 Meson", *Phys. Rev. Lett.,* vol. 13, no. 4, pp. 138-140, 1964.
[http://dx.doi.org/10.1103/PhysRevLett.13.138]

[9] M. Ahmadi, B.X.R. Alves, and C.J. Baker, "Characterization of the 1S-2S transition in antihydrogen", *Nature,* vol. 557, no. 7703, pp. 71-75, 2018.
[http://dx.doi.org/10.1038/s41586-018-0017-2] [PMID: 29618820]

[10] W. Ubachs, "ArXiv 1511:00985v1 [physics.atom-ph]",

[11] M. Kajita, *Measurement, Uncertainty and Lasers.* IOP Expanding Physics, 2019, pp. 1-8.
[http://dx.doi.org/10.1088/2053-2563/ab0373]

[12] H. Fischer, "A history of the central limit theorem (from classical to modern probability theory)", *Springer,* 2021.HistoryCentralLimitTheorem.pdf (mcgill.ca)

Conclusion

In this book, I summarize the fundamental part of analysis for the whole field of physics. Until 19th century, the fundamental of whole physics was classical mechanics, giving the equation of motion under a certain potential field. The electromagnetism gives the formula of the potential field, given by the electric and magnetic fields.

In the 20th century, the limit of validity of classical mechanics was found. Quantum mechanics is required to describe the phenomena on the micro-scale. The fundamental of quantum mechanics is wave-particle duality. The wave characteristics of particles were derived from the analogy of the characteristics of light, which were derived from Maxwell's equation. The quantum mechanics converges to the classical mechanics by $h \rightarrow 0$ (h: Planck constant).

The theory of relativity is required to describe the phenomena with high velocity. The theory of relativity was based on the constant speed of light in the vacuum c, which was derived from Maxwell's equation. The theory of relativity converges to the classical mechanics with $c \rightarrow \infty$.

Therefore, all fields in physics are closely correlated. It is most important to understand the role of each field to establish the whole physics. For example, knowledge of classical mechanics is required to understand quantum mechanics. On the other hand, a simple imagination of quantum mechanics makes it easy to learn classical mechanics more in detail.

I am now writing while the world is in panic with the outbreak of coronavirus. But seeing that Newtonian mechanics was established during the plague outbreak, we might consider it as the chance to find our next step.

SUBJECT INDEX

www.ingramcontent.com/pod-product-compliance
Lightning Source LLC
Chambersburg PA
CBHW041711210326
41598CB00007B/608